Advanced Lighting Controls:

Energy Savings, Productivity, Technology and Applications

Advanced Lighting Controls:

Energy Savings, Productivity, Technology and Applications

Edited by Craig DiLouie

THE FAIRMONT PRESS, INC.

CRC Press
Taylor & Francis Group

Library of Congress Cataloging-in-Publication Data

Advanced lighting controls : energy savings, productivity, technology and applications / edited by Craig DiLouie.

p. cm.

Includes index.

ISBN 0-88173-510-8 (print) -- ISBN 0-88173-511-6 (e-book)

1. Electric lighting--Automatic control. 2. Electric power--Conservation. I. DiLouie, Craig, 1967-

TK4169.A38 2005

621.32--dc22

2005044905

Advanced lighting controls: energy savings, productivity, technology and applications/ edited by Craig DiLouie

Published by The Fairmont Press, Inc.
700 Indian Trail
Lilburn, GA 30047
tel: 770-925-9388; fax: 770-381-9865
http://www.fairmontpress.com

Distributed by Taylor & Francis Ltd.
6000 Broken Sound Parkway NW, Suite 300
Boca Raton, FL 33487, USA
E-mail: orders@crcpress.com

Distributed by Taylor & Francis Ltd.
23-25 Blades Court
Deodar Road
London SW15 2NU, UK
E-mail: uk.tandf@thomsonpublishingservices.co.uk

Printed in the United States of America
10 9 8 7 6 5 4 3 2 1

0-88173-510-8 (The Fairmont Press, Inc.)
0-8493-9863-0 (Taylor & Francis Ltd.)

Table of Contents

Preface

Lighting controls are an essential part of every lighting system and a major frontier in building and energy management.

An estimated 30-45 percent of a building's electricity bill is typically represented by the cost of operating lighting systems. And 30 percent to 35 percent of the cost of a building is for the mechanical systems and envelope architecture. Automated lighting controls can contribute significantly to cost savings in these areas.

According to the New Buildings Institute, which developed the *2001 Advanced Lighting Guidelines*, automatic lighting controls can reduce lighting energy consumption by 50 percent in existing buildings and at least 35 percent in new construction. In addition, lighting automation has proven effective in load shedding and peak demand reduction, resulting in additional direct cost savings in addition to potential incentives from utilities with demand response programs. Numerous strategies and technologies are available so that a proper combination can be matched to individual application needs.

Besides energy management, benefits of lighting automation include mood setting via the ability to alter a space through dimming or color changing; flexibility by allowing users to instantly adapt a space to different uses; ability to establish a responsive lighting system that can be globally and locally controlled, with automatic operation; ability to adapt electric lighting systems to daylighting strategies; decrease "light pollution" (skyglow, light trespass and glare) by dimming or switching lights based on time of night or occupancy; enhancement of workspaces with a technology that has visible effects; and potential increased worker satisfaction by enabling users to control their own light levels. The list goes on.

Lighting automation can be completely automated or contain elements of manual operation; can be localized, global or both; can be hardwired or wireless; and can be used for automatic switching or dimming. A wide variety of proven and developing technologies is now available to achieve a wide variety of building and energy management goals. New approaches, such as the Digital Addressable Lighting Interface (DALI), light fixtures integrating automatic controls, and control of LED lighting systems, offer new opportunities while existing technologies continue to develop in capabilities, interoperability, ease of specification and use, and reliability. New developments such as LEED, demand response programs, changing workplace goals, rising energy costs and the ASHRAE/IES 90.1-1999 (or later) energy code continue to stimulate demand for lighting automation. Research indicates that lighting automation is becoming the norm, not the exception. Both the use of automatic switching controls and dimming controls are increasing.

Advanced Lighting Controls was developed to help construction and building management professionals view lighting automation from a number of angles. It is intended as an introduction to the technology and surrounding technical, legislative and related issues and opportunities. A majority of the content for this book was written by the editor with input from the members of the Lighting Controls Association, a non-profit organization dedicated to educating the industry about the benefits, operation, technology and application of lighting automation. Members of the Lighting Controls Association include Advance Transformer, HUNT Dimming, Leviton Manufacturing, Lightolier Controls, Lithonia Lighting, Lutron Electronics, OSRAM SYLVANIA, PCI, Square D, The Watt Stopper, Tridonic and Universal Lighting Technologies.

Advanced Lighting Controls provides significant background to help construction and building management professionals consider lighting automation as an effective energy and building management strategy.

Section I

LIGHTING CONTROL

Chapter 1

Introduction to Lighting Control

By the National Electrical Manufacturers Association, Lighting Controls Council

Lighting controls have gained an extraordinary degree of popularity in recent years because they pay for themselves so quickly due to the energy savings and other benefits they can provide.

The demand for controls created by their rapidly growing popularity has encouraged manufacturers to invest millions of dollars in research and development, to bring to the market new controls that are even more versatile, more reliable, and more cost-effective than ever before. In fact, modern lighting controls tend to create clear and convincing evidence that a building is up to date, by relying on technology that has been expressly designed to enhance the flexibility of lighting while at the same time avoiding waste.

It is particularly interesting to see what has happened to the cost of lighting controls over the years. While the price of so many other products has increased, the cost of modern lighting controls has come down, due in large part to the twin impacts of mass production of electronic components and competition. At the same time, the value of the benefits associated with lighting controls—energy savings, demand reduction, increased productivity, and more retail sales, to mention a few—has risen steadily.

LIGHTING CONTROL FUNCTIONS

Lighting controls perform seven discrete functions: on/off, occupancy recognition, scheduling, task tuning, daylight harvesting, lumen depreciation compensation, and demand control. Some lighting controls perform only one function; many perform more than one, typically on an automated basis. The following discussion provides more detail about each of these functions.

On/Off

The basic control function, typified by the common wall switch, is turning lighting on or off. The degree to which this function is performed depends on other variables or control functions such as occupancy recognition and scheduling, which are described below.

Occupancy Recognition

Occupancy recognition is commonly used in intermittently occupied areas or rooms, typically to turn lights on when people are present and off automatically after a certain amount of time when they are no longer present. Experience indicates that occupancy detection can save significant amounts of energy and money by preventing the waste caused by keeping lights on when they are not needed.

Contemporary occupancy recognition devices rely on one of two principal technologies: *ultrasonic* or *passive infrared.*

Ultrasonic systems transmit an inaudible sound in the frequency of 20,000 to 40,000 Hz to a receiver. Any movement alters the transmitted sound waves and is recognized by the receiver, causing it to initiate control action.

Passive infrared sensors use a pyroelectric detector and a fresnel lens to sense the radiation emitted naturally by people. Movement of the "heat source" is transmitted through the lens to the detector, triggering a control event.

Occupancy recognition is "packaged" into a variety of systems. In some, they serve only to turn lights off, in case the individual leaving the room forgets to. In others, they are used in combination with dimming equipment, to raise illuminance when a person approaches—e.g., at a display case in a lightly traveled area of a store, and, later to lower illuminance to the predetermined point

Scheduling

When scheduling is applied, electric illumination in given areas is activated, extinguished, or adjusted according to a predetermined schedule. In some cases, the systems control may be vested in a different device. For example, the system indicated in Figure 1-1 would be under the direction of daylight harvesting controls from 9:00 am through 4:00 pm and, from 11:00 am to noon, and 2:00 pm to 4:00 pm, demand management controls would have precedence.

Scheduling is a time-based function and, as a consequence, it is most suited for facilities or spaces where certain things happen at certain times. Because "off-normal" conditions inevitably arise, local overrides usually are provided.

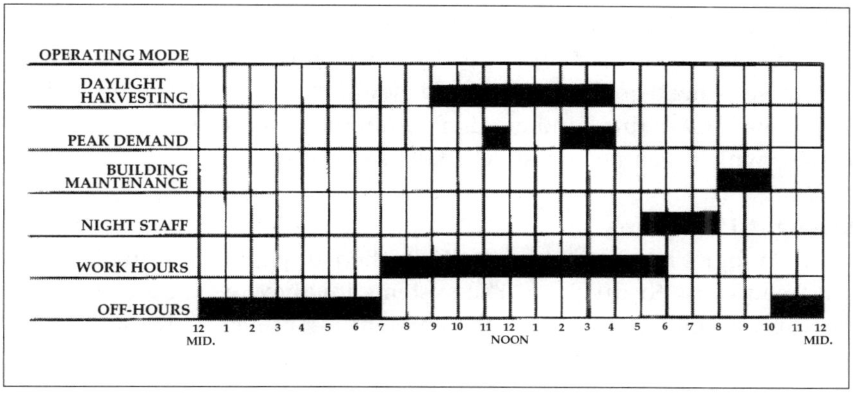

Figure 1-1. Typical weekday lighting schedule.

Tuning

Tuning means adjusting the light output of a luminaire or a system of luminaires to the specific level needed for the task or other purpose, such as aesthetics. It is most commonly done through dimming. It can also be accomplished through switching, as when ballasts of a four-lamp luminaire are wired in such a way that the two inboard and two outboard lamps are separately switched, permitting full light output or 50 percent light output.

Tuning can create significant monetary benefits through energy use reduction. In essence, it helps assure that only the amount of light needed is actually provided. The more flexible and easily controlled the system is, of course, the more benefits that can flow. For example, when a given worker is able to adjust electric illumination to optimal levels for that specific person, productivity will be higher. Long-term benefits are also apparent, as when tasks change or are relocated. Rather than having to move luminaires or replace them altogether, it often is possible to meet new needs simply by changing light output.

In retail areas, the ability to provide tuning creates the ability to define spaces with light, to create a mood or atmosphere most suited to the nature of the display, and to highlight impulse purchase items or seasonable goods.

Tuning is also used for aesthetic purposes, when light output is adjusted to create dramatic effects of one type or another.

Virtually any type of lighting system can be tuned, and particular advances have been made in the field of fluorescent and high-intensity discharge (HID) lighting. While dimmable fluorescents have been available for many years, new control modules and electronic ballasts now help assure high-quality effects and new levels of cost savings. Similarly, new HID fixtures and auxiliary equipment enable light levels to be varied, which was not possible a few years ago.

Daylight Harvesting

Daylight harvesting is applied when daylight entering a space can't be put to positive use. The systems involved use strategically located photocells to determine the ambient light level. This information is fed to a control device that then raises or lowers luminaire output or turns off selected luminaires to maintain the amount of light (illuminance, measured in footcandles) set for the space. The adjustment occurs gradually, so occupants in a space are not aware of it. Response delays are also used to prevent frequent adjustments due to passing clouds or similar phenomena.

Some buildings are designed to take advantage of daylight. Others have daylight available to them and using that daylight may or may not be worthwhile, depending on factors such as the tasks being performed and/or the orientation of workstations with respect to windows. Daylight also brings heat with it, which, in summer, might necessitate cooling unless appropriate window films are installed. In other words, if a building has not been designed to use daylight, some study is needed to help assure it can be put to positive use and to establish exactly what needs to be done in order to realize that gain.

Lumen depreciation compensation. The output of electric illumination systems diminishes over time, due particularly to a phenomenon called lamp lumen depreciation (LLD). As shown in Figure 1-2, most commonly used lamps produce less light the longer they are in service. Light also is lost due to the build-up of dust and dirt on lamps and the reflective surfaces of luminaires, as well as other reflective surfaces in the illuminated spaces, including walls and ceiling.

Lumen depreciation compensation is essentially the same as daylight harvesting. It senses ambient luminance and increases light output to maintain whatever is desired. At such time as the desired illuminance cannot be provided, lighting system maintenance is called for, and probably is overdue.

Although effective controls can help compensate for inadequate maintenance, they should not serve as an excuse for poor maintenance. In fact, regular maintenance of electric illumination systems can often be a major source of energy and cost savings, as well as improved lighting quality.

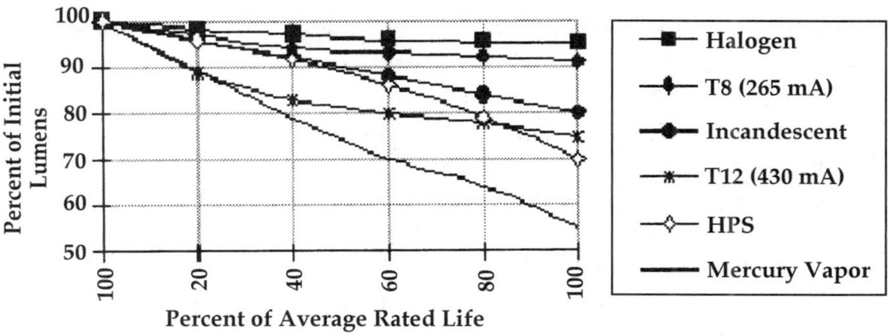

Figure 1-2. Lamp lumen depreciation of commonly used lamps. *Source: Lighting Design Lab.*

Demand Control

Most nonresidential electrical rate schedules impose a charge for energy and demand. Energy equates to the amount of kilowatt-hours consumed in any given billing period. Demand is the rate at which energy is consumed. The more energy needed at any given moment in time, the more the utility must do to provide it—i.e., more generating capacity, more distribution capacity, and so on. To illustrate by example, consider two hypothetical buildings. Both consume the same amount of electricity, but building "A" consumes it in 24 hours, building "B" in eight hours.

Even though both buildings consume the same amount of energy, the utility obviously must invest more money in generating, transmission and distribution equipment to meet the needs of building "B." The extent of this investment is defined by the highest rate at which energy is consumed, even though it may be consumed at that rate for only a very short period of time.

Energy charges alone would yield a poor rate of return on the utility's investment because the utility's equipment needed for building

"B" is used only for a relatively short period of time.

One way of obtaining a more reasonable return would be to average all generation, distribution, and transmission equipment costs among all customers. That approach would not be fair, however, because those who use the utility's equipment efficiently would be subsidizing those who do not. To be fair, therefore, the utility collects from each customer an amount proportional to the cost of meeting that customer's demand requirements. Demand requirements are measured through the use of special metering equipment which measures and averages consumption for a certain period of time. This period is called the *demand interval*.

A commonly used demand interval is 15 minutes. The typical demand meter records average energy consumption for each 15-minute interval in a day. When the first interval ends, the equipment resets and starts on the second one.

The utility reviews the demand records for the building at the end of each billing period. The maximum demand recorded is used to compute the demand charge.

Demand charges can be substantial. Demand control equipment is used to help assure that demand will not exceed a given maximum. The procedure involves identification of certain secondary loads that can be "shed" during peak periods. In some cases, it may mean that air conditioning is shut down for a given period of time, then restarted once demand ebbs somewhat. Certain lighting circuits also can be made part of the secondary loads, with some luminaires being dimmed while others may be turned off altogether. Typical candidates would be lobby lighting systems, overhead office lighting, and other electric illumination systems that can at least be dimmed for short periods without creating any adverse impact on safety or security.

Demand control becomes particularly important when a utility has what is called a "ratchet clause" in its demand schedule. In essence, a ratchet clause states that the amount of demand for which a customer is billed should reflect the maximum demand recorded at any time in the recent past, since the utility must be prepared to meet a customer's demand requirements at any time, not just during a given month.

Most utilities experience maximum demand in summer months due to the widespread use of electric cooling systems. A typical ratchet clause, therefore, may state that the amount of demand for which a building is billed during the winter period may be no less than a certain percentage of the maximum demand recorded during the previous summer season. If a 75 percent ratchet clause is in effect, and if maximum summer demand was 1,000 kW, the winter season demand bill would be based on a minimum of 750 kW (1,000 kW x 75 percent). Even if the actual winter season demand never exceeds 600 kW, the demand used for billing purposes will be 750 kW. Obviously, if demand control equipment is used to keep demand as low as possible during periods of peak use, savings will be achieved for all subsequent periods to which a ratchet clause is applied.

Electric utilities throughout the United States are encouraging more demand control, since every kilowatt of demand that is reduced adds a kilowatt of new capacity. By eliminating waste, America's utilities can continue to meet new demand requirements without having to build costly new generating plants. For this reason, many are subsidizing the cost of lowering demand by offering rebates and other financial incentives. Lighting controls can play a vital role in this important undertaking.

THE BENEFITS OF LIGHTING CONTROLS

Modern lighting controls provide an array of benefits, ranging from energy savings and electrical demand reduction to supports of the functions for which lighting is needed. The bottom-line value of some of these benefits can be significant, creating paybacks that are best measured in weeks rather than years.

Energy Savings

Controls are the only devices that can help assure optimal use of energy and elimination of energy waste. By applying controls wisely, a building owner or manager can help assure that only the specific amount of lighting actually needed, if any, is provided. No matter how much efficiency may be designed into a system through selection of lamps, luminaires, ballasts, and shielding/diffusing media, maximum energy efficiency cannot possibly be achieved without effective controls.

Utility Cost Savings

For some people, utility cost savings and energy savings mean one and the same, but that seldom is the case. Almost all electric utilities impose a demand charge on other-than-residential usage, and the cost involved can be substantial. Through effective controls, utility costs can be reduced through reduction of energy consumption (measured in kilowatt-hours or kWh) and demand (measured in kilowatts, or kW).

Increased Worker Productivity

The importance of lighting controls to worker productivity is underscored by the fact that maximum worker productivity cannot be attained unless workers are given optimal lighting conditions; that is, that quantity and quality of light that are best suited to the nature of the tasks involved and an individual worker's visual capability. If all tasks and all workers' eyesight were the same, lighting controls would not be needed to help maximize productivity. But the visual needs of workers vary considerably, and the tasks they perform not only differ, but can be altered during a given workday. Workers need to be able to adjust their lighting in order to create the conditions they need to perform their work as quickly and efficiently as possible. Tuning fills this need.

It is worth nothing, too, that many of today's office tasks rely on electronic equipment that uses a video display terminal (VDT) as the person/machine interface device. Controls can be highly effective in reducing VDT screen glare—i.e., reflections in the VDT screen that can obscure alphanumeric and graphic displays. This glare requires operators to work at a slower pace and under high stress. Left uncorrected, VDT screen glare and associated problems can reduce workers' productivity and increase the frequency and severity of their errors, thereby decreasing the overall cost-effectiveness of workers in an organization.

In a test conducted by the Illuminating Engineer Research Institute (IERI) to determine the impact of different lighting levels on subjects' ability to proofread mimeographed documents, it was found that the number of errors made decreased as light levels were improved. It was also shown that older workers made more errors than younger workers, and that improved lighting had a far more pronounced impact on error reduction for older workers than for their younger counterparts. See Figure 1-3.

This study, as many others, points out that older workers, often considered among the most reliable employees, need better quality lighting to offset physiological changes that affect their eyes due to the aging process. While the reduced errors and improved productivity that result from better lighting can justify greater lighting expense, the gains often are realized with reduced lighting expense, through proper selection of lighting management options.

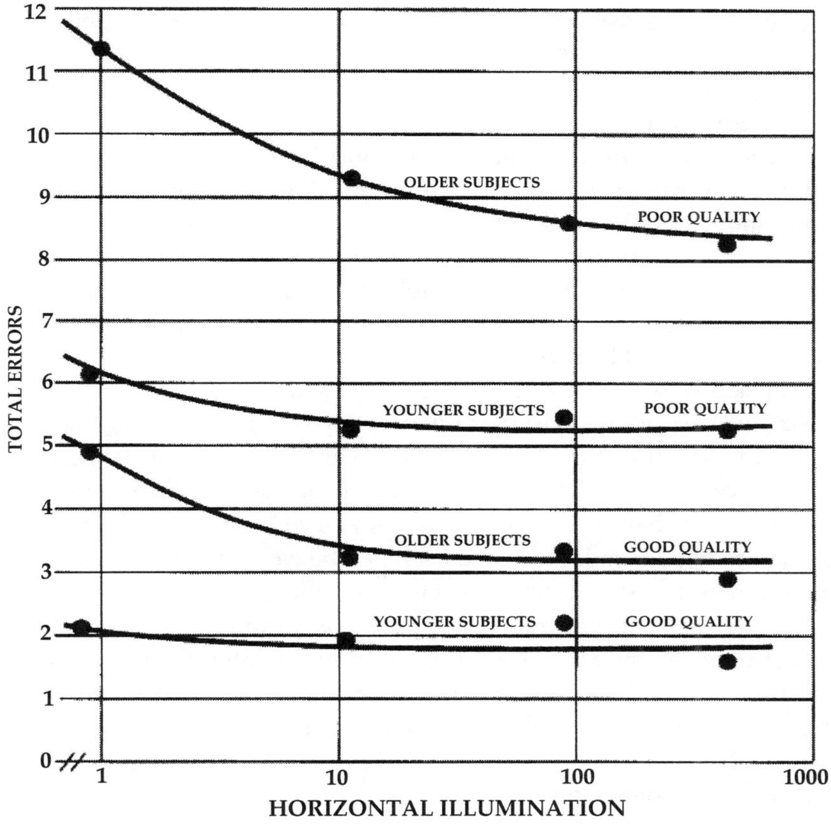

Figure 1-3. Results of an IERI study show that frequency of errors declines when illumination levels are increased.

Pollution Prevention

Lighting controls offer the potential of saving electricity which, in turn, prevents air pollution caused by electricity generation. Carbon dioxide (CO_2) is a greenhouse gas that contributes to global warming—35 percent of all CO_2 comes from electric utilities. Sulfur dioxide (SO_2) is the major contributor to acid rain—65 percent of all SO_2 comes from utility sources. Nitrogen oxides (NO_x) turn up in our environment as smog and acid rain—36 percent of all NO_x comes from electric utilities.

Consider this example: If a standard light switch is replaced by an occupancy sensor in a room with four fluorescent fixtures (192 W per four-lamp fixture), the sensor could prevent 1,505 lbs. of CO_2, 11 lbs. of SO_2 and 6 lbs. of NO_x from being released into the environment each year.

Error Reduction

The same lighting control that contributes to enhanced worker productivity can also contribute to error avoidance. The cost of errors can be huge, leading to lost time and, potentially, more serious consequences. Particularly in those areas where an error can have costly consequences, obtaining the best possible lighting is a wise investment.

Expanded Space Flexibility

It has been reported that office layouts are modified on the average of once every three years. If the space is to be flexible, lighting must be flexible. Failing that, the lighting system would have to be overhauled every time the space is rearranged or different tasks are introduced to it. By having a lighting system whose luminaires' light output can be easily adjusted, the space itself can be easily adjusted, to accommodate new tasks and/or workstation locations. The cost savings involved can be immense, by avoiding the cost of luminaire relocation or replacement, or by avoiding the even more significant costs that can result when people are forced to perform their work in a space where lighting is a detriment rather than a support.

Improved Aesthetics and Image

Lighting controls can have a significant impact on lighting's ability to enhance aesthetics and affect image. Indoors, for example, lighting can be used to create highlights and contrasts, establish visual effects on walls, illuminate specific objects, and otherwise affect the appearance of a space and the objects in it. Lighting's abilities in this respect can be fine-tuned only with controls. Controls are needed in order to keep specific lighting designs effective by compensating for lamp lumen disparities. And, when objects are moved or replaced, controls are needed to adapt the existing lighting to its new tasks. Controls are also valuable in areas affected by daylighting, to maintain visual interest and appeal as the angle and intensity of sunlight changes. Controls are the answer when it comes to creating varying lighting levels or even to alternating lighting systems created specifically for aesthetic purposes, as when the color of the illumination moves from reds, to pinks, to whites.

Outdoors, much the same is true. Through lighting controls, a system can gain a level of sophisticated and impact that otherwise would not be possible. And despite the many different ways in which lighting can be altered to such significant effect, the cost of the controls needed is small when compared to the value of impacts such as more effective marketing, better curbside appeal, enhancement of prestige for tenants, and more noticeability and attraction.

Mood Setting

Lighting affects mood and, through use of controls, those who control lighting in a space are in a position to affect the moods of those using the space. Restaurants or other dining areas comprise a typical case in point. To create an air of intimacy and romance, lighting can be "turned down low." In conference rooms, a low level of illuminance can be used to encourage a relaxed approach to the topics at hand, while a far higher illuminance might be used to stimulate people.

Better Space Marketability

Many control benefits are well-known. As such, many of today's office managers or other responsible for locating new or additional space are looking for effective lighting as part of the package. When space is separately metered, so that individual tenants pay their own utility bills, it is important for them to have lighting controls to help keep those bills as low as possible. Controls also are essential to help adjust the lighting to meet the task and aesthetic needs involved. And, by having modern controls in place, it should not be necessary to spend much, if anything at all, to make the existing lighting fully compatible with whatever tasks a new occupant will be performing, no matter where in the space those tasks are performed. In short, effective controls contribute to the marketability of space.

Space Savings

In many circumstances, it may be necessary to have two or more spaces in order to support functions that have markedly different lighting needs. Through controls, it is possible to support multiple functions in one space, by being able to select exactly which luminaires will be activated, when, and to what extent. In the case of the Chesapeake & Potomac Telephone Company in Washington, DC, for example, one space was used for both a conference room and a video conferencing broadcast studio. Originally, designers thought that two spaces would be required. By using controls effectively, initial costs were reduced significantly, as were long-term costs.

Heightened Security

Lighting controls play a significant role in safety and security applications. For example, occupancy sensors can be used for daytime lighting control and for after-hours use, so that all lighting or a series of lights is activated instantly in the event of a detectable intrusion. Another example is a card access (entry/exit) system that can provide a command to the lighting control system to turn the lighting on for an occupant entering the building after hours. If the occupant turns on other lighting zones, the security personnel might be alerted to check the area.

More Effective Facility Management

Microprocessor-based lighting controls can make a facility far more responsive to the needs of building management personnel by effectively monitoring tenant lighting energy usage and costs. In some cases, a historic comparative analysis of lighting energy cost by individual load can be performed to identify operating problems. These types of controls can also contribute to better maintenance, by compiling lamp runtime and cycle data, basic factors that determine when maintenance is required. This permits less lighting equipment downtime, which increases tenant goodwill and permits performance of maintenance operations.

Improved Worker Morale

Better lighting often causes an improvement in employee morale, not only because the new lighting often is more comfortable, but also because it enhances the appearance of the illuminated space. Employee morale can be affected even further when employees can have individual control over their lighting, because it permits them to convert the lighting into an individualized tool. It also gives employees more control over their own space.

Environmental Enhancement

When individual customers reduce their energy demand and use, a utility can serve more people from the same generating facility. Over time, as customer efficiency is enhanced, less electricity has to be produced per capita, resulting in fewer pollutants being discharged into the air, especially by utilities that rely on coal. While the actions of one person or the energy performance of one building may not have much impact on the environment, using that as an excuse to put off positive action no longer is acceptable to many Americans. Problems such as acid rain are frighteningly real and pose serious concerns for the future. As such, if our future is to have a more secure environment, each person and each building must make a contribution, by not being a source of waste. Without effective control of lighting, waste is inevitable. This waste is harmful to the nation's environment and to the pocketbook of whomever must underwrite it.

LIGHTING CONTROL OPTIONS AND THEIR APPLICATIONS

Many types of lighting controls are available to permit "real time" management of electric illumination. With few exceptions, these controls are applicable to virtually all types of buildings and most types of spaces—e.g., offices, stores, restaurants, hotels and motels, hospitals, warehouses, prisons, factories and museums.

Lighting controls generally can be categorized as *manual* or *automatic*.

Manual controls turn lighting systems on or off, or adjust light output, in direct response to manual adjustment—e.g., flicking a switch or moving a dimmer slide. Manual lighting controls include lighting panelboard controls (circuit breakers) and contactors for controlling large numbers of luminaires, wall switches for flexible control for small groups of fixtures, key-activated switches for applications where lighting control security is important, and solid-state manual dimmers.

Automatic controls are either programmed to take a certain action at a specified time, or the action is event-initiated. Examples of automatic controls include time-based programmable controls for indoor and outdoor switching, photocell controls that respond to changes in light levels, occupancy sensor controls that operate by sensing the presence of people, and microprocessor-based programmable and network control systems that provide flexible lighting systems control and integration.

For purposes of this discussion, control options are grouped as switching controls, dimming controls, and integrated lighting control systems. Each is discussed below.

Switching Controls

Switching controls turn lights on and off, and many perform other functions as well. At a minimum, every space should be equipped with manual switching to permit occupants or facility management to control lighting usage. Switching off lighting when it is not needed not only reduces lighting energy consumption, it also results in less lighting system heat build-up, reducing the cooling load and air-conditioning needs. This reduces energy consumption further and provides additional savings. The following discussion addresses switching controls in ascending order of control intelligence.

Lighting Contactors

Lighting contactors permit manual or automatic control of large blocks of lighting loads. Three types of lighting contactors in common use are: feeder-disconnect-type (rated up to 1200A to control large blocks of load); multipole contactors with as many as 12 poles (rated 20A) for multibranch circuit control; and single-pole relays rated 20A with low-voltage control for individual branch circuit or luminaire control. Contactors are used with many forms of automatic controls, as through integration with solid-state lighting control modules that operate as a function of photocell or occupancy sensor input or with microprocessor-based energy monitoring and control systems.

Local Wall Switches

Local wall switches (AC snap switches) are the most commonly used control devices for local lighting control. They can handle a full 20A branch circuit lighting load—e.g., 24 to 26 four-lamp fluorescent fixtures at 277V.

For best results, switches should located convenient to users, to encourage deactivation of lighting whenever appropriate, to reduce energy waste. Wall switches also can be applied to develop a flexible lighting control scheme. As an example, consider Figure 1-4 which represents one section of an office illuminated by 12 four-lamp fixtures.

Two wall switches can be used, one to control all outboard lamps (A and D) while the other controls the inboard lamps (B and C). Alternately, one switch could be used to control all even-numbered fixtures and another for the odd-numbered, or to control fixtures 1/2, 5/6 and 9/10 separate from the other six. Any of these techniques permits a 50 percent reduction in light output, with the best selection being that which closely matches occupancy patterns in the space. By using four controls, even more variations are possible.

Key-activated Switches

Key-activated switches are wall switches that turn lighting on and off by a key. They are installed to prevent unauthorized or accidental use of certain lighting circuits. They are particularly useful for HID light sources that must cool down before they can be activated.

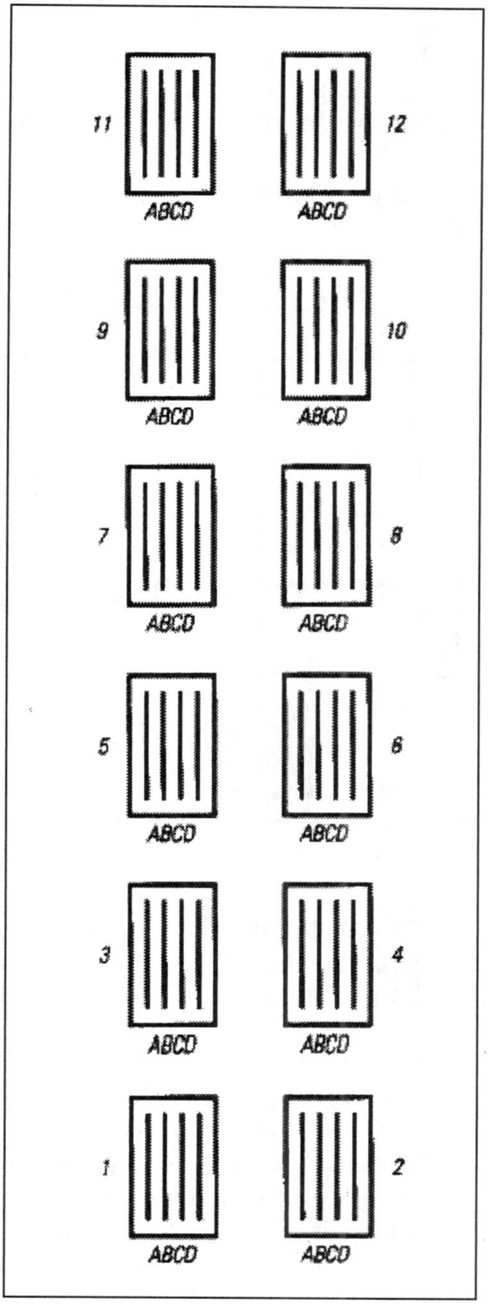

Figure 1-4. Layout of 12 four-lamp fluorescent fixtures.

Intelligent On/Off Local Devices

Intelligent on/off local devices consist of at least two elements: a logic or intelligence module and a power switching device. The logic or intelligence elements vary depending on the needs of the specific applications.

Figure 1-5 illustrates two intelligent on/off local configurations that provide a simple approach to controlling a single load, using only one intelligent input.

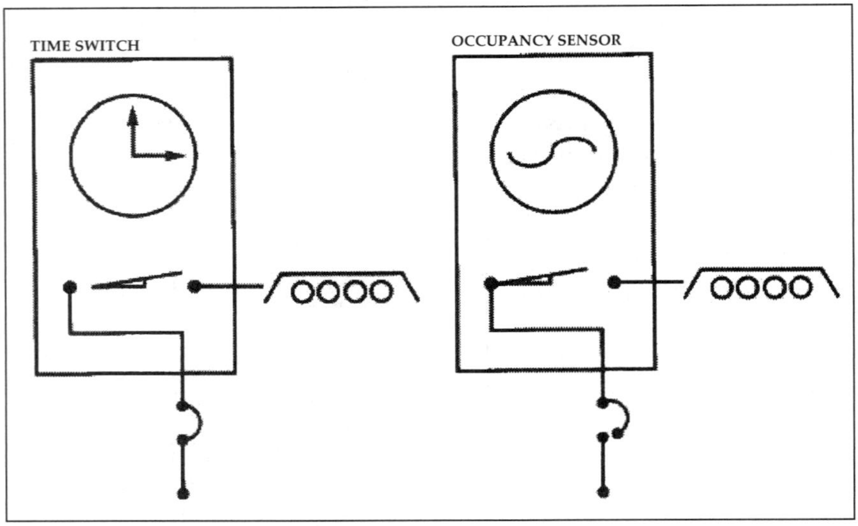

Figure 1-5. Examples of on/off local configurations.

The intelligent input in its simplest form can be a time control or an occupancy sensor. Each typically is used to control a single load and is wired directly to it.

Time Controls

These controls, also known as time clocks or time switches, activate and deactivate their loads at user-determined times. They are available as electromechanical or electronic devices.

Many types of electromechanical time controls are manufactured. The 24-hour time switch is a basic unit, usually capable of activating and reactivating a load at least several times each day. Used indoors, it could activate lighting at 8:00 am, deactivate it at noon, reactivate it at

1:00 pm, and deactivate it again at 6:00 pm. Through the use of two 24-hour time switches, selected luminaires could be turned off and on at different times or, through split ballasting, different lighting levels could be obtained at different times each day. A seven-day time switch affords the same daily selections as the 24-hour device for one week at a time.

Figure 1-6 illustrates a time switch with an astronomical feature that automatically compensates for sunrise/sunset time shift during the year. This option provides an alternative to photocells for controlling exterior lighting.

Most time switches are available with back-up drives to maintain accuracy despite a blackout or brownout. Some rely on spring-wound mechanisms, others use batteries. Both 24-hour and seven-day time switches are available with a day-skipping feature. This keeps selected loads off during holidays and weekends.

Microprocessor-based time controls provide a higher degree of flexibility than electromechanical devices, permitting users to program more on/off actuations each day, and to create special schedules for

Figure 1-6. Time switch with astronomical feature.

holidays or certain functions.

Indoors, time control is ideally suited for those applications with high predictable occupancy, such as stores and factories. When used in an application with nonpredictable occupancy such as an office, the time function should be supplemented with local overrides. In such cases, the time control should be capable of multiple offs to provide "off sweeps" to catch those areas overridden and left on. Electronic time controls are needed for that purpose.

Care must be taken not to put an occupied area in total darkness with a timeclock off. Multiple-level switchings to provide various lighting levels can be employed to avoid exposing occupants to a safety risk.

The most advanced time controls automatically flick lights off and on to warn occupants that their area is about to go off, and then protect the individual overrides from the next timed sweep. The time control's intelligence must be able to recognize an override by the occupant while the lights are still on and protect that override. These occupant-sensitive scheduling devices may provide a timed override with another warning when the override is about to go off.

To assure proper application and occupant convenience, indoor time controls could be evaluated using the following criteria:

- Ability to provide on/off actuation matched to the needs of the space or load;

- Ability to be overridden by a local switch with automatic return to the schedule mode;

- Ability to maintain the operating schedule in the event of a power outage; and

- Ability to provide a warning in occupant areas that the lights are about to go off and then protect the occupant override.

Occupancy Sensors

Occupancy sensors (see Figure 1-7) are automatic switches that control lighting based on the presence or absence of people. Their primary function is to switch electric illumination off automatically in an unoccupied space after the last person leaves that space. A timing control provides light for a period of time after the area is vacated. Some models offer variable control while others have a fixed time delay. Issues such as reduced lamp life, because of frequent switching may enter the decision of using sensors. In most cases, the reduced hours of lamp operation and energy savings more than offset any effect the switching may have on lamp life.

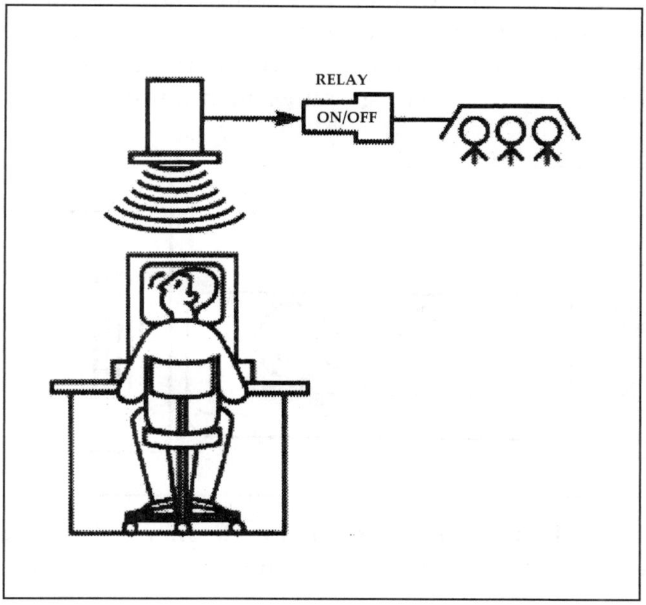

Figure 1-7. Occupancy sensor system.

Repeated tests have shown that single-person offices are occupied only about five to six hours a day. Nonetheless, lights in these offices and other intermittently occupied areas often are left on for as many as 14 hours. In such ceases, occupancy sensors can easily achieve energy savings of 30-50 percent.

Various types of occupancy sensors are available for mounting on the ceiling or in a wall box. The ceiling-type usually operates a small control unit that contains a relay. The sensor sends low-voltage pulses to the relay, which then switches the controlled luminaires on or off. The type designed for wall-box mounting usually is sized for installation in a standard wall switch electrical box. Some models are capable of interfacing with integrated microprocessor-based lighting controls to provide additional control capabilities when occupancy is detected.

As already discussed, most occupancy sensors rely on ultrasonic or passive infrared technology. The ultrasonic devices shown in Figure 1-8 permit two user adjustments. One of these makes the device more or less sensitive to motion. In spaces where personnel remain relatively quiet and stationary, such as a library, more sensitivity is required to prevent inadvertent deactivation of lighting. The second adjustment determines the amount of time the control will keep lighting on after no motion is detected. Passive infrared occupancy sensors are so designated because they detect energy rather than transmit it.

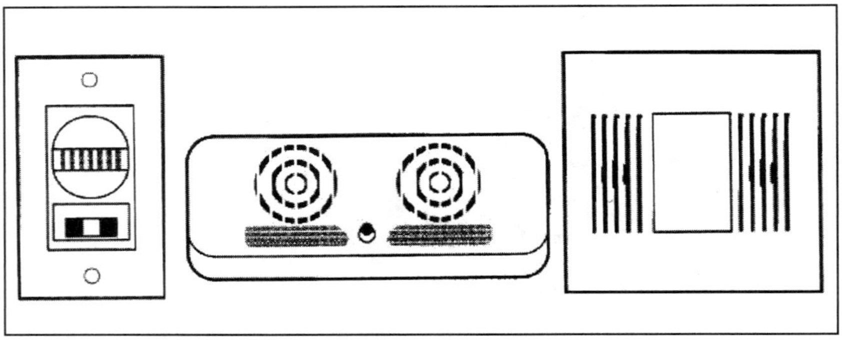

Figure 1-8. Wall- and ceiling-mounted occupancy sensors. Sensitivity and activation delay adjustors are located inside the sensor.

Occupancy sensors may be mounted on a ceiling or wall. Although they have no maximum area limitations, the high partitions found in open office areas limit their coverage and may necessitate use of additional sensors.

In a typical installation, relays and transformers are connected to the sensors by low-voltage wire. Alternatively, all relays can be installed in one location for operation through a master controller. The master controller contains the power supply for the individual sensors, the timing adjustment control (permitting independent sensor-by-sensor settings), and an override switch to permit local, manual lighting control.

Occupancy sensors can be integrated with others such as a dimmer. When no motion is detected, lighting would be kept on, but at a predetermined low level. Then, when motion is sensed, lighting would be brought to a higher level or full output.

Some manufacturers of occupancy sensors suggest that their products can be used to operate both lighting and certain types of HVAC equipment (terminal units and multizone system dampers) and provide interface modules for that purpose. Occupancy sensors can also be used for intrusion protection with or without an audible (local and/or remote alarm).

A wide array of sensors is available. These include sensors that replace standard wall switches, using the same box and wiring, to flush-mounted and surface-mounted units that are designed for wall and/or ceiling placement, to those used specifically for halls and stairways, or for outdoor security lighting systems. These vary considerably with respect to the amount of square footage covered and overall "field of view."

The ability of ultrasonic and infrared sensors to detect minor motion and to avoid false activation in unoccupied areas varies considerably among the various products presently available. To help ensure occupant convenience and economic practicality, consider the following criteria in selecting a unit:

- Detection of minor motion. In order to avoid any inadvertent deactivation of lighting in occupied areas, the sensor should be sensitive to people movements such as turning a page in a book, picking up a telephone, or shifting in a chair.

- Large area coverage. The return on investment will be affected by the number of sensors required to cover a given area.

- Installation requirements. Units that are easy to install will reduce costs and improve the return on investment.

- Appearance. When sensors are visible, they should be attractive enough to help assure acceptability by occupants of the building.

Photocells

Photocell controls respond to changes in ambient light. When the ambient light level falls to a user-determined level, lighting is switched on. When the ambient light increases to a user-determined level, lighting is switched off.

Insofar as outdoor safety and security are concerned, a photocell control is superior to a time control because it can respond to overcast conditions during daylight hours. Most photocells are equipped with a delay feature (at least one minute) to prevent the rapid on/off cycling that otherwise could occur during a partly cloudy day.

Indoors, photocells are being used for daylight harvesting and lumen depreciation compensation. Some systems rely on one strategically placed photocell to operate all appropriate fixtures, with manual switches usually being installed as overrides. Others use one photocell per luminaire with the photocell being aimed directly below the fixture or over the most critical process or surface requiring illumination. The best system is that which is designed to meet the specific needs of the facility involved, in the most cost-effective manner. As such, in some areas, fixtures may each have one photocell, while in other areas or zones it may be best to rely on just one photocell to control several fixtures.

Another concept in photocell control involves the use of two photocells: one indoors and the other outdoors (or pointing outside through a window). Data from the two photocells is transmitted to a control package that consists of a differential amplifiers, an on/off three-minute delay timer, and an output relay. When the outdoor pickup senses more light than the indoor pickup, the control lamps are de-energized. They are activities again when daylighting diminishes. Sensitivity controls are available for fine-tuning.

Photocells also can be used to create multiple-level switching. In Figure 1-9, for example, each two-lamp ballast controls one lamp in the luminaire where the ballast is housed, and one lamp in the adjacent fixture. Thus, when the ambient lighting level falls below the predetermined minimum, all lamps are activated. When the ambient lighting level increases to a predetermined point above the minimum but below the maximum the "A" ballasts (or "B" ballasts) are activated creating 50 percent lighting output. At such time when the maximum ambient level is achieved, all electric illumination is deactivated. A similar approach can be accomplished through reliance on multiple-level ballasts, where code permits.

A solid-state phototiming device also is available for outdoor applications (Figure 1-10). The device is an astronomical minicomputer programmed by sunlight sequences and driven by a microprocessor. Using dawn and dusk as a reference, the device calculates actual solar time to operate its control functions. Included in the self-programming logic are daylight patterns during daylight savings time, allowing the device to calculate and reset to daylight savings time within five days.

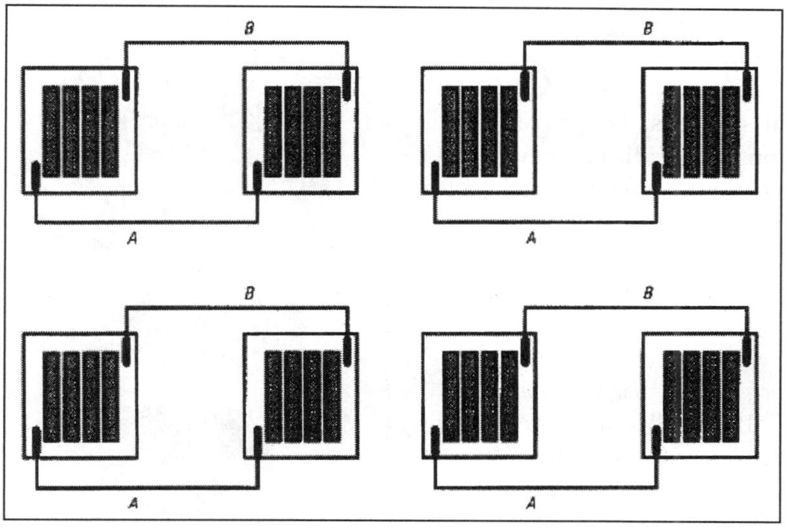

Figure 1-9. Photocell-based multiple-level switching layout.

Low-voltage Controls

Low-voltage switching systems provide a more flexible switching platform than standard line voltage switches. The simplest system consists of a transformer that produces 24V or less, relays that are wired to the loads, and on/off switches that are connected by low-voltage wiring to the relays (Figure 1-11). Each relay can control up to a full branch circuit (20 amps). Low-voltage wiring provides inherent wiring flexibility while also providing the foundation for simple lighting automation.

Low-voltage switching often is used to solve complex switching problems. In particular, it allows any number of switches to be used to control a single load. This simplifies central and local control of lighting from several locations, pilot lights provide status indication. Because small low-voltage cables replace line voltage wiring and conduit, this type of remote switching becomes economically viable.

Local (Figure 1-12) and remote master switches can be added to allow master control of a floor or department and still allow an individual to override local lighting. Timeclocks or building automation then can be used to control the lighting automatically while still allowing an individual to override a particular area for after-hours use. Timed "sweeps" catch the overrides.

As shown in Figure 1-13, existing keypad phones can be used instead of (or in addition to) local switches. This saves installation labor while providing a convenient switching method for occupants. It is particularly effective in open office areas or for the control of remote loads or buildings.

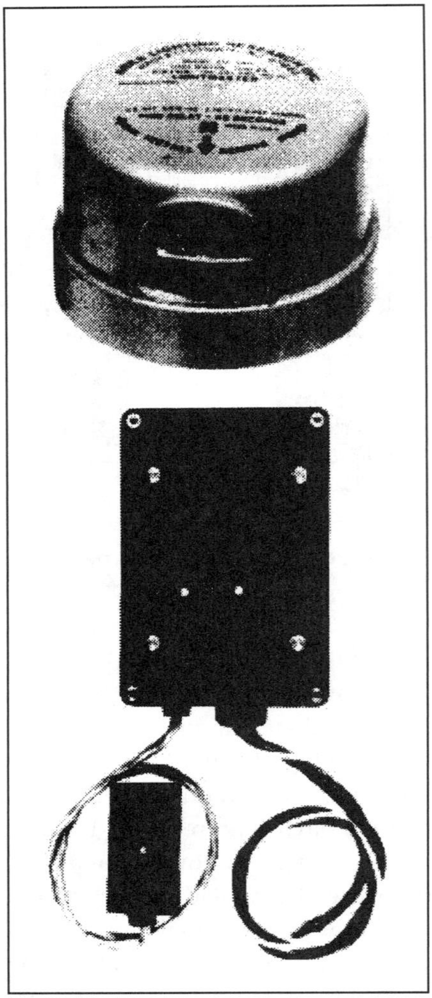

Figure 1-10. Solid-state phototiming device.

Power-line Carrier Controls

Power-line carrier control systems create an alternative to extensive re-wiring in retrofit or the installation of control wiring in new construction. These systems rely on small receivers installed inside luminaires to control ballast operation. Transmitters send coded command signals to these receivers over existing electrical wiring. Some of the transmitters are wall-mounted and look much like conventional wall switches; others are centrally located. Most power-line carrier ballast load-switching systems are centralized and use toggle switches or similar devices to effect operator control. Many of these systems also can be operated automatically through microprocessor-based devices or electro-mechanical time controls.

Two-level HID Controls

Two level HID lighting controls are relay systems that operate mercury vapor, metal halide, and high-pressure sodium lighting at ei-

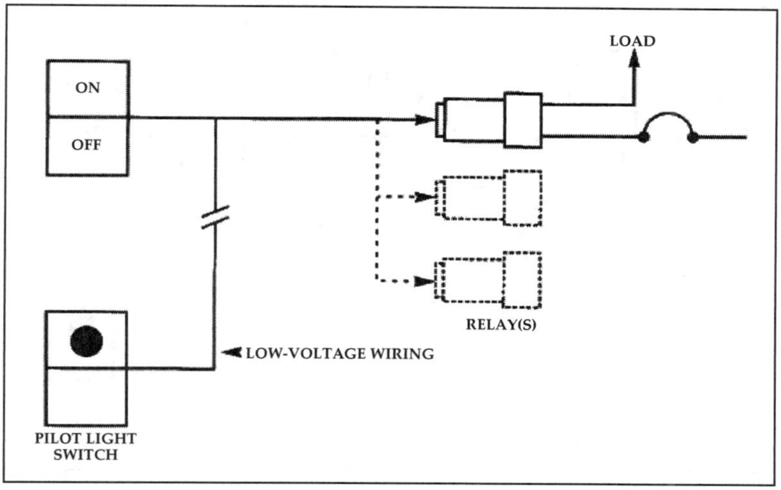

Figure 1-11. Low-voltage control system schematic.

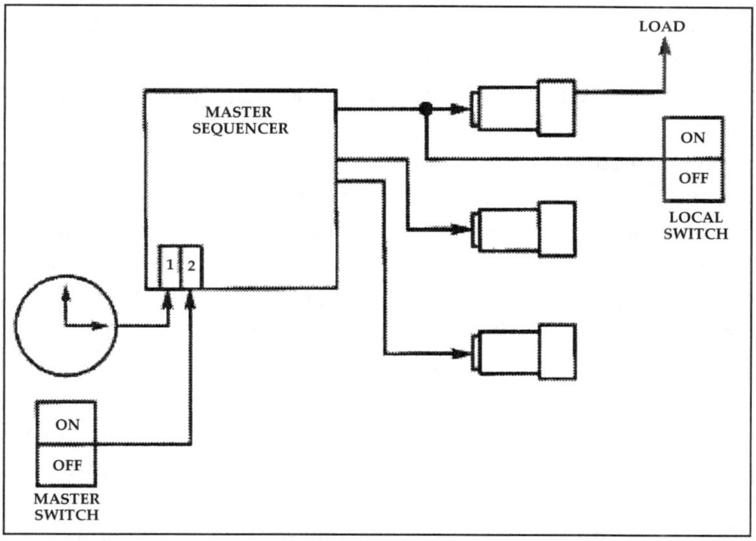

Figure 1-12. Lighting automation with local switch override.

ther full light output or less (e.g., 50 percent). System components include an on/off contactor, remote switching ballasts for operation of HID luminaires (also available as a factory-installed option), and various control equipment, such as photocells, occupancy sensors, and time

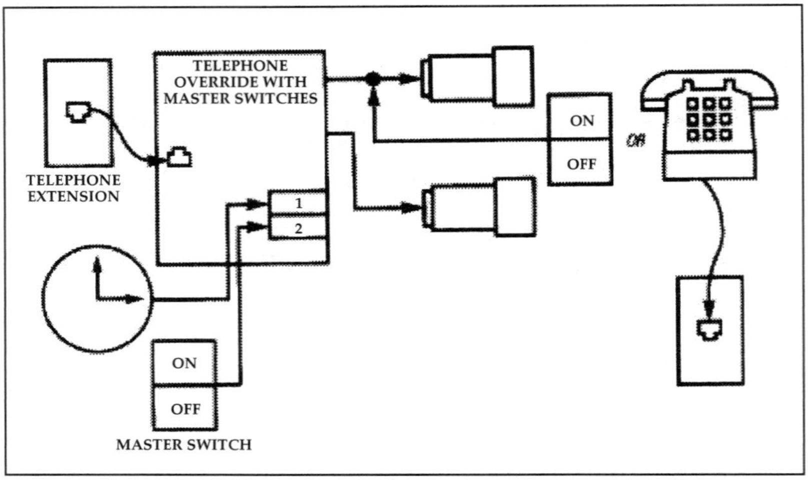

Figure 1-13. Lighting automation with telephone override.

clocks (Figure 1-14). Available for lamp wattages from 150W to 1500W, these systems are well-suited for use with airport aprons, warehouses; prisons/security facilities, parking lots, and ship-loading docks.

Dimming Controls

Dimming controls are available for most types of lighting. They can be integrated into automatic lighting control systems and can be used manually as well. Some dimming controls require use of magnetic or electronic dimming ballasts, while others employ an electronics package installed in the panelboard or elsewhere within the system.

Dimming control technologies typically rely on either *voltage reduction* or *waveform management*. Voltage reduction is used principally with incandescent lighting. Full-range dimming is obtained by lowering the line voltage to the lighting systems without significantly affecting the shape of the AC line voltage. This technique is applicable to low-voltage incandescent lighting as well, except some solid-state transformers may not be capable of full-range dimming. Although voltage reduction also can be used with gas-discharge lighting (fluorescent, mercury vapor, metal halide, high-pressure sodium, and low-pressure sodium), its effectiveness is limited unless special dimming ballasts are used.

Waveform management systems effect dimming by modifying the shape of the AC line voltage. Phase control is the most common type of

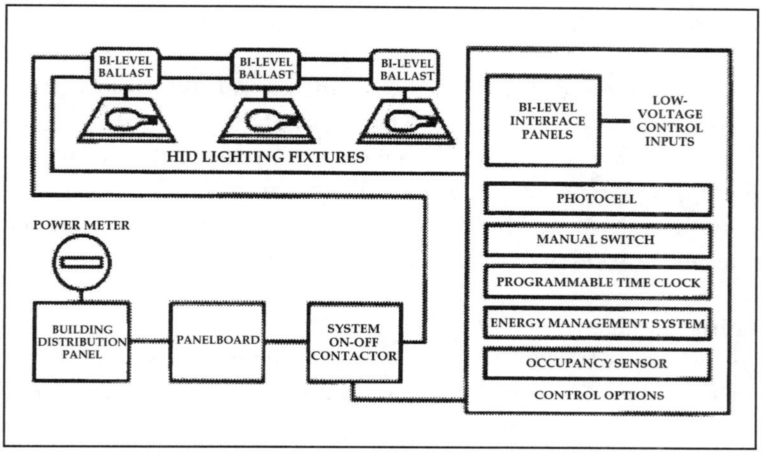

Figure 1-14. Schematic of two-level HID lighting control system.

wave form management. It operates by eliminating the initial portion of each half-cycle of the AV waveform. The amount of dimming achieved is determined by the amount of each half-cycle that is eliminated. Typically, these devices permit 25 percent to less than 100 percent of nominal incandescent light output (100 percent output usually cannot be obtained due to a small voltage drop across control elements). Phase control can also be used with gas-discharge lighting providing special magnetic dimming ballasts are applied. In such cases, fluorescent lighting output can be controlled from 5 percent or less to full nominal output.

Phase control dimming equipment is available in a variety of shapes, sizes, and functions. Wallbox controls are available with rating from 600 to 2000W; larger modular system dimmer packs can handle any load. Control schemes can include any of the time clock, photocell, or other arrangements already mentioned, as well as a wide range of manual controls and an ability to interface with building automation equipment. Even wallbox dimmers, which have generally been considered stand-alone devices, now can be interfaced and controlled by external systems.

Other types of waveform management controls permit dimming of gas-discharge sources without use of special ballasts, but the low end of their output range is somewhat high, from 15 percent to 50 percent, depending on the specific device employed. Dimming controls' low initial cost makes them attractive for retrofit as well as new construction.

Although these newer controls are not yet available in wallbox size, they can be obtained as: fixture-mounted devices that control a single ballast; subcircuit devices that control up to eight ballasts; and 20- to 100-amp circuit control devices. They are available as stand alone units, with manual or photocell control, as well as units designed for integration with other building control systems.

Dimming Devices

A number of dimming devices are available. These can be categorized as wallbox dimmers, integrated dimmers, modular dimmers, low-voltage dimmers, preset dimmers, and variable output ballasts, as follows.

Wallbox Dimmers

Wallbox dimmers are manual controls that give occupants more control over their visual environment. They often are applied in high value areas, e.g., executives offices and multi-purpose rooms such as audiovisual training or presentation areas.

Various control configurations are available, including those that use linear slides, rotary knobs, raise/lower buttons, preset panels, and even wireless remotes.

Linear slide and rotary dial dimmers are available in 600W and 2000W models for various types of lighting: incandescent, low-voltage, florescent, cold-cathode, and neon. Some of these units also are provided with buttons that activate lighting to a present level.

In the case of low-voltage incandescent lighting systems, both single-pole and three-way low-voltage dimmers are specified. Three-way low-voltage dimmers are used with standard three-way switches; dimming is possible from one location only. Architectural-style low-voltage dimmers also are available for higher-rated lighting loads. Several manufacturers offer rotary, slide control, and preset slide dimmers for commercial applications.

Integrated Dimmers

Integrated dimmers integrate a variety of features into a wallbox configuration. Commonly included features are: multiple channel control, where all or selected luminaires on a circuit are controlled by a single dimmer; multiple presets; and universal circuitry that allows each dimming channel to control incandescent, low-voltage incandescent, fluorescent, cold-cathode, or neon light sources.

Integrated dimmers can be categorized into two levels of control. Level I integrated dimmers control a single circuit and offer multiple channel control with either a single preset or multiple presets. Level II integrated dimmers control multiple circuits and multiple channels, with either a single or multiple presets. These presets permit various lighting scenes to be created in several lighting zones throughout the building. As an example, one control permits four-scene preset dimming control in a single wallbox. Auxiliary controls, such as infrared wireless transmitters and activators, provide remote location switching control of all or any single scene. In a commercial application, the system could create unique lighting scenes to enhance a conference room for a variety of functions, such as lectures, presentations, slide projections, or meetings. Other commercial applications include private offices, restaurants, lobbies, museums, and shops. The system also can be applied in residences. For example, lighting scenes could be preset in a dining room for breakfast, lunch, dinner, and entertaining. Preset scene also could highlight a piece of art or illuminate a specific entrance at different times of day.

System Dimmers

System dimmers offer lighting control for larger applications where wallbox and integrated products are impractical or higher performance is required. These systems consist of dimmer cabinets and control stations, typically connected with low-voltage control wires as shown in Figure 1-15. Cabinets may contain any number of dimmers capable of handling small and large loads controlling a variety of lamp sources. Control stations can range from a simple manual slide control to multiple preset controllers in a variety of configurations. Typical applications include churches, restaurants, meeting rooms, and multiuse facilities.

Variable Output Ballasts

Variable output ballasts represent the latest trend in dimming control technology. In essence, waveform dimmer circuitry is made part of an electronic fluorescent ballast. When interconnected to a photocell or photocell-based system (e.g., multiple photocell inputs to a micropro-

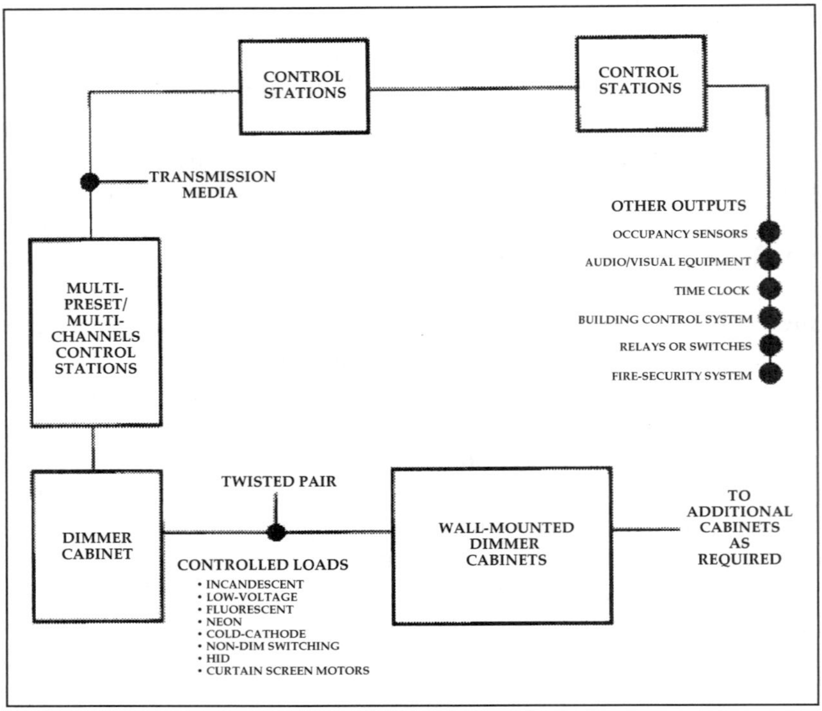

Figure 1-15. System dimmer.

cessor that transmits signals to the ballast), the light output of the lamp or lamps controlled by the ballast is adjusted in a predetermined manner. These systems are ideally suited for daylight harvesting lumen depreciation compensation, and load-shedding for demand control.

Daylight Harvesting Controls

Although day-lighting controls can be implemented with on/off systems, they are far more effective when dimming is incorporated. Most of the various dimming technologies make provision for daylight control schemes, and even those that do not can review daylight information from photocells included in a building automation system. Since periods of maximum daylight harvesting potential correspond with periods that experience maximum air conditioning demand, daylight-based lighting controls can limit peak energy demand as well as save large amounts of energy. The actual energy savings to be achieved depend heavily on specific application factors such as glazing area and design illuminance. In most cases, savings of 30 percent or more of the lighting energy used in daylight-controlled areas can be achieved. If the entire space is uniformly skylighted, these savings can accrue on the entire lighting load. More commonly, they apply only to the perimeter zone of a vertically glazed installation.

Note that the amount of energy to be saved by a daylight harvesting system does not increase dramatically when the minimum output level available from the dimming system is less than about 25 percent. Because of their greater dimming range, waveform control systems and variable output ballasts are well suited to daylight harvesting applications. Also, since some systems may be less efficient at the low end of the dimming range, it is important to compare the power at the low end, not just the light level, when evaluating these controls.

Generally speaking, two types of daylight harvesting needs exists, distinguished by the distribution of daylight in the controlled area. Perimeter zone applications are the most common since daylight enters a space through vertical windows. The distribution of daylight tends to be highly nonuniform, with large amounts in areas close to windows and rapidly decreasing amounts further away. In these situations, it is desirable to control luminaires adjacent to the glazing separately from those further in to obtain maximum energy savings while still providing necessary task illumination. Waveform control dimmers or variable output ballasts are well suited for the application, because they can be sized to the appropriate control zone. Depending on the dimming system chosen, it may be on the dimming system chosen, it may be best to specify that power wiring for the luminaires run parallel to the windows rather than radically outward from the building core. This can be an important consideration in retrofit or renovation installations.

The second type of daylighting situation generally occurs in skylighted areas where the distribution of daylight is relatively uniform throughout the controlled space. Large-area waveform control gear or voltage reduction equipment may be well suited for such applications, although the limited dimming range of voltage reduction gear may be too restrictive in some cases.

Perimeter-zone applications are more design-sensitive; proper photocell selection and placement are critical. Several techniques are being employed to help assure that photocell input is proportional to the amount of light on tasks in the space. No single technique is best for all applications; each application must be evaluated individually, and for that reason, designers should ask manufacturers to supply a calibration procedure for the particular type of application.

On/off switching of interior lighting as a function of available daylighting is inexpensive but intrusive. For this reason, photocells should provide switching at "safe levels" with a wide deadband and built-in time delays to avoid nuisance switching. Also, because daylight differs from electric lighting in color and directions, it usually is recommended that the electric illumination not be switched off entirely. Typically, daylight area are split-wired and switched to the 50 percent level during periods of adequate daylighting. If outdoor luminance is measured as the basis for switching, local overrides should be provided for individuals who close their blinds.

To achieve maximum savings and end-user acceptance, daylight harvesting systems should incorporate these features:

- Window film or treated glass to inhibit the closing of blinds and drapes due to the sun's glare and heat; an automatic shut-off to turn lights off after 6:00 PM; and

- Timed override controls to turn lights back on by people who are working after 6:00 PM.

Lumen depreciation compensation controls: Any of the dimming technologies can be used to provide lumen depreciation compensation, with the choice being based mostly on the size of the area and type of source. When considering the use of lumen depreciation compensation controls, be sure to evaluate various group relamping strategies to find the combination that provides the best results for given application. The proper choice helps assure good lighting along with low energy and maintenance costs for the life of the lighting system.

Be cautious about combining daylight and lumen maintenance functions. Although it seems to be an easy and obvious way to achieve extra savings, since a photocell is required for both, achieving satisfactory results is not as easy as it seems. Lumen depreciation compensation photocells generally are high-gain devices compared to those used for harvesting daylight. A high-gain photo-cell generally will create excessive dimming with only a small amount of daylight present, while the low-gain daylight harvesting photocell generally will ignore variations in light level due to lamp or direct depreciation, particularly for spaces with highly nonuniform distribution of daylight. In cases where highly uniform daylighting is provided, the lumen deprecation compensation sensor can provide adequate daylight compensation as well, but these situations must be regarded as more the exceptions than the rule. As such, attempting to use one photocell for both functions may compromise the other function severely.

Integrated Lighting Control Systems

Integrated lighting control systems (Figure 1-16) consist of manual or automatic components designed to control compatible electronic dimming ballasts. They can be interfaced with other microprocessor-based centralized lighting control systems or building automation systems. These systems can perform all of the functions that are critically important to energy optimization. They can sense conditions in each area or zone and control lighting to yield maximum energy efficiency without affecting visual comfort or other conditions. Some of the newest, most advanced systems use distributed processing, resulting in "smart" components that have their own microprocessors permitting them to operate effectively without having to "ask" the central control unit for instructions.

Data collected from various input sensors and commands issued to various remotely controlled points are sent through field interface devices and from the central control unit via data transmission media. Some of the more commonly used transmission media are:

- *Twisted pair wiring* is used to create a dedicated hardwired line between the lighting controllers sending or receiving data. Its transmitting performance is similar to that of a coaxial cable, as are its expandability and maintenance requirements.

- *Coaxial cable* consists of a center conductor surrounded by a shield that protects against electromagnetic interference. Coaxial cable can operate at data transmission rates that are limited only by the data transmission equipment. Its multiplexing capability means there is no practical limit on the number of facilities that can be connected to the system, making it an excellent choice, especially when expandability is a concern.

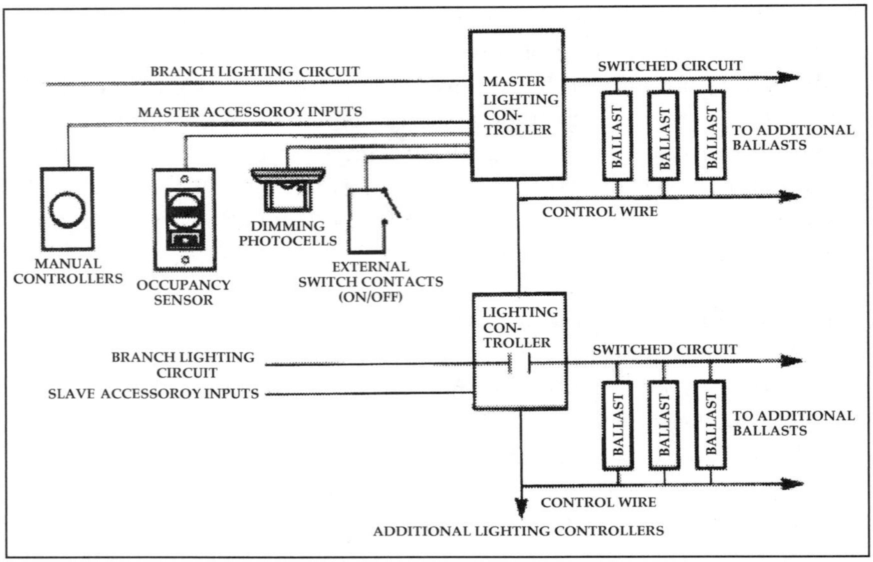

Figure 1-16. Integrated lighting control system.

- *Triaxial cable* has the same characteristics as coaxial cable. Composed of a coaxial cable plus an aluminum-mylar outershield and drain wire, it is used where the cable will not be run in conduit.

- *Fiber-optic (FO) cable* uses the wideband properties of infrared light traveling through transparent fibers. It is best suited for point-to-point high-speed data transmission. The signal attenuation of high-quality fiber optic cable is far lower than the best coaxial cable. Repeaters are required for every 2,000 feet of coaxial cable, but are three to six miles apart in fiber optic systems.

- *VHF or FM radio signals* used for start and stop functions are popular, but problems have been experienced obtaining frequencies and, when they are obtained, interference can occur. Expandability also can be a concern, along with limited signal distances, high maintenance requirements (due to the large number of transmitters and receivers involved), and low reliability. A combined system is sometimes used, whereby an FM radio signal and a carrier signal are carried on a power line.

- *Microwave transmission* is a practical alternative for communication between facilities separated by considerable distances. Microwave transmission affords fast scan rates, excellent reliability (assuming knowledgeable maintenance personnel are available), and compatibility with future requirements and expansion. The primary problem is high first cost: receivers and transmitters are needed in each building.

- *Telephone lines* are the most commonly used data medium when the lighting control computer is remote from the building(s) served. The local telephone company charges a small initial connection fee and ongoing fees for monthly equipment lease. Maintenance is included in the monthly lease fee, with a certain level of service guaranteed.

- *Power-line carrier signals* can be used to transmit data to remove locations within the building complex using carrier current transmission that superimposes a low power radio frequency (RF) signal, typically 100 kHz, onto the 60 Hz distribution system. Since the RF carrier signal cannot operate across transformers, all communicating devices must be connected to the same power circuit, or RF couplers must be installed across transformers and receivers to be connected over a wider area of the power system.

Microprocessor-based centralized programmable lighting control: A microprocessor-based centralized programmable lighting control system is basically a microprocessor-based centralized controller. Although it is designed principally for lighting, it is capable of handling other loads.

Photocells can be integrated into the system, as well as split-ballasting controls. Other possibilities include demand control, duty-cycling, and computer integration. The system can also handle HVAC, service water heater, and motor loads.

Microprocessor-based programmable controllers can be integrated into networked lighting control systems that allow schedules and other programmable functions to be entered and then modified from a central operator console. Networked systems also allow input from sensing devices such as master switches, photocells, occupancy sensors, telephones or load-shed contacts to control relays or dimmers, thus reducing wiring costs. In addition, the network allows the central collection of operating data and status information for building management functions.

Networked lighting control systems communicate with each other and with a central terminal (usually a personal computer) utilizing a variety of transmission media, as discussed above. In order for the network to differentiate between devices, each must have a unique address or identification.

Three basic types of networked lighting control systems are used: *polled, interrupt* and *tokens*. They provide several three incremental functions: 1) central programming and monitor/control, 2) global commands, and 3) management data.

The typical networked control system shown in Figure 1-17 provides cost-effective automated lighting control for applications ranging from a small office building to a mall to an industrial complex. Each of the distributed control panels has stand-alone automation capability. The network links these controllers to a central operator terminal (PC).

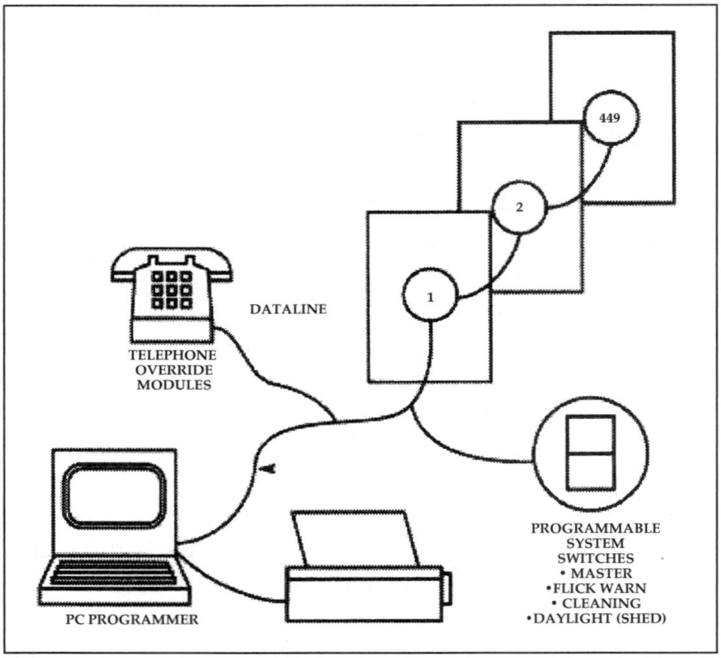

Figure 1-17. Networked intelligent panels.

Besides supporting such features as telephone control and distributed master switching, these systems excel in providing management data. For example, at the end of the month, the operator can simply ask the system for a report on the total lighting energy consumption for the last period. If that consumption is excessive, the operator then might ask for a report of every load that exceeded its expected runtime during the month. Having identified the "offenders," a profile of the actual runtime for each be used to identify why and how the excess occurred. Such management data is critical to ensuring that automated lighting systems continue to save energy. In addition, this same information can become the basis for a fair allocation of lighting costs by tenant or department.

Providing networked lighting controls also ensures that the lighting can be effectively integrated with other building controls to provide full intelligent lighting operation. For example, the card access system can be linked to the lighting to turn on all associated hallways, work areas, cooling and fans when an employee comes in on a weekend. From the lighting perspective, an occupancy sensor tied into a network can not only turn on the local lighting, but this information can be relayed to the grand station to provide security information for the building.

Building Automation Systems

Building automation controls generally are microcomputer- or mini-computer-based systems that are capable of controlling lighting systems as well as HVAC, security, and fire safety systems. Depending on the options specified, they can perform many other functions, too, such as maintenance scheduling (in a variety of ways), monitoring, logging, and inventory control. In fact, it has been stated that we have only begun to realize the many different functions that computerized systems already available can perform.

The approaches used for lighting control are essentially similar to those associated with microprocessor-based centralized programmable control systems. Lighting systems can be integrated easily and virtually all the different functions described above can be controlled from one central location, relying on the appropriate sensors, actuators, and monitors, connected together by multiplexed data transmission media.

Section II

DESIGN AND PLANNING

Chapter 2

How to Design a Lighting Control Scheme

By the Lighting Controls Association

The three main steps to creating a successful lighting control design and seeing it through are:

1. Conceptual design
2. Final design
3. Construction observation

CONCEPTUAL DESIGN

Elegant lighting design addresses the whole building, the site, and the occupants as an integrated system. The lighting designer must take into account such factors as color, form, space, emotional connotations, patterns of use, and much more. It's not a trivial cookbook process, but the rewards are well worth the effort.

Successful projects usually result from good communication between all parties, and clear objectives. The design team—architect, lighting designer, engineers, etc.—must work closely together and with the owners during the whole design process to be sure that the design goals are clearly understood by all. It may be helpful to develop a formal checklist of required, desired, and not allowed factors.

These are the steps in the conceptual design process for a successful, integrated lighting controls design:

- Encourage and envision the daylighting
- Present the lighting controls as a part of a greater philosophy
- Understand the building and its occupants
- Identify lighting control opportunities
- Perform a conceptual economic analysis

- Get the support of other team members
- Get the client excited about lighting controls

Encourage and Envision the Daylighting

Sunlight is beneficial to both physical and emotional comfort. Although substantial savings can result from using daylight, the benefits to the occupants from exposure to healthy sunlight are much more significant. A daylighted building should need only minimal electric lighting during daylight hours, especially in sunny regions. Lighting controls can be used to dim or turn off electric lighting when bright sun makes the electric lighting unnecessary. This can result in substantial savings, due to the reductions in both power demand and energy use. They can also help "blend" the electric lighting in with the daylight, for a smooth transition from daylight to electric light as the daylight level decreases.

For the health and well-being of occupants, encourage the use of sunlight and understand the look and feel of available sunlight in the building based on the building's orientation, and the locations of glazing, light shelves, etc. Look for opportunities to reduce the amount of electric lighting with daylight-driven dimming or on/off controls.

Present the Lighting Controls as a Part of a Greater Philosophy.

A systems approach and integrated design are better than a piecemeal approach. Encourage the integration of the lighting and lighting controls with the architecture, the available daylight, and the environmental controls systems. Integration can lead to money savings as well as a sophisticated and simplified end result.

Understand the Building and its Occupants

Get to know the building. Is it a new building or a renovation project? What types of spaces and ceiling heights will be found in the building? What is the approximate square footage of each area?

It is also essential that the design team understands the interests, lighting needs, expectations, and behaviors of the occupants. The designer needs to know the tasks that the occupants will be called upon to perform (and the visual difficulty of performing those tasks); the occupants' work schedules, and the likely pattern of use for each space and for the lighting in those spaces. Interviews or user surveys of the actual occupants can be very helpful. If this isn't possible or practical, use survey data for similar groups in the same geographical area.

Good data on occupants can have an impact on the weighting given to design factors. (For example, a 20-year-old needs one-third less light than a 60-year-old for the same task.) The owners will be better able to make sure that the result is satisfying to actual users, not just aesthetically pleasing or under budget.

Identify Lighting Control Opportunities

Identify general areas or percentages of the building for which certain kinds of lighting control may be suitable. It may be useful to color-code the various possible control schemes on a building floor plan (i.e., daylighting and occupancy controls in the open offices, manual dimming in the private offices and centralized controls for the entire building). Consider the relevant past experiences of the owners and team members and know the budget available when contemplating the right level of complexity for the conceptual control scheme.

Sometimes it's appropriate to keep the overhead lights on even if daylight levels are very high in the area. Controls can dim the electric light so it appears to still be on, but the lights are not consuming wasted energy.

Employees may be annoyed if their overhead lighting is turned off, even if daylight levels are high. This is especially true if an indirect lighting scheme is used, because the ceiling will be considerably darker if this type of lighting is turned off.

Dimming is one of the best solutions to this sort of dilemma. An automatic daylight-driven dimming system can dim the lights down to 20 percent, 10 percent, or even down to 1 percent when daylight levels are high. Even when dimmed to such a low level, fixtures appear to be on, making the store feel "open for business," and making the ceiling of the office space bright.

Perform a Conceptual Economic Analysis

You'll need to present the conceptual control design (i.e., color-coded diagram or list of control ideas) to the client and/or the owners, and they are bound to wonder about the bottom line. Perform a simple, rough economic analysis for the conceptual design. Note that 30 percent to 45 percent of a building's electricity bill is typically for lighting. And 30 percent to 35 percent of the cost of a building is for the mechanical systems and envelope architecture. Lighting controls can contribute significantly to cost savings in these areas.

Get the Support of Other Team Members

Lighting controls can have beneficial effects on other areas of the building design. Start early to convince fellow team members of the benefits of lighting controls. If everyone understands that the controls are an integral part of the design, it's less likely that the controls will be cut from the project further on in the process. For example, if the use of lighting controls in the design allows first-cost savings in the HVAC system, then the controls could pay for themselves instantly.

Get the Client Excited about Lighting Controls

Take the opportunity to discuss lighting controls with the owner, who stands to benefit the most from their use. Not only will there be economic benefits, but the quality of the building as a whole will be higher and the occupants could be happier and more productive due to the personal choice and added flexibility. Several benefits are expressed in Figure 2-1.

FINAL DESIGN

This is the step in which specific lighting and lighting controls products are selected and located on the plans. These are the primary goals to accomplish during the final design phase:

- Provide a reliable, correctly-operating system
- Provide lighting flexibility where it is needed
- Design a system that is convenient to use and to maintain
- Satisfy the occupants
- Reduce the needed capacity of the HVAC system
- Minimize energy consumption
- Satisfy security needs
- Bring the project in on time and within budget

The main steps in the actual design process are:

Space Type	*Benefit*
Discount Retail Store	In an open retail space with daylighting, dimming can reduce electric lighting use but allow the lights to be on, making the store seem "open for business."
Conference Room, Classroom, Auditorium, etc.	Dimming lighting can facilitate a variety of visual presentations.
Health Care Facility	Daylight-driven dimming can provide a smooth and unnoticeable transition to electric lighting as daylight levels decrease, while maintaining the desired light level.
Restaurant	Preset scene dimming controls can make changing the ambiance as the day goes on consistent and as easy as pressing a button.
Office Area	Even in an open office area, occupants can be given the option of dimming the light fixture over their workstation to suit their personal preferences.

Figure 2-1.
Benefits of automatic lighting controls in various space types.

- Design controls for each area
- Compile construction documents

This is the step during which controls are in the most danger of being cut. See the Kansas City New Zoo project in the Case Studies Appendix for an example of how this can happen.

Design Controls for Each Area

The first step is to systematically evaluate all the parameters involved in the design in light of the design goals. For each area, you need

OCCUPANCY AREA	*ENERGY SAVINGS*
Private Office	13 - 50 percent
Classroom	40 - 46 percent
Conference Room	22 - 65 percent
Restrooms	30 - 90 percent
Corridors	30 - 80 percent
Storage Areas	45 - 80 percent

Figure 2-2.
Typical energy savings with occupancy sensors.
Source: U.S. Environmental Protection Agency.

to determine which components will be most appropriate. At the same time, you need to decide on the optimum placement for each component. Seek assistance from the control manufacturer. Many controls manufacturers are more than willing to help make sure you've selected the appropriate devices for each area, that your control scheme will work. They'll also give you wiring diagrams to give to the contractor; most of them will do it for free.

Select Products

Depending on the relative importance of the several factors, select appropriate components and test the integrated design to see if it will satisfy the goals.

Control Voltage

Some controls can be hooked to line-voltage power, and others must be connected to low-voltage (DC) power. For a new building, both are possibilities, but for a partial renovation, it may be beneficial to use low-voltage controls.

Typical coverage patterns (applies to occupancy sensors). There are several different kinds of coverage patterns and mounting configura-

OPERATING COST COMPARISON
PRIVATE OFFICE, 128 SQ. FT.

Performance	*Base Case*	*Occupancy Sensors*	*Daylighting*	*Occupancy Sensor + Daylighting*
Annual Energy Use[a]	450 kWh	340 kWh	330 kWh	250 kWh
Annual Energy Cost	$33	$24	$24	$18
Annual Energy Cost Savings	—	$9	$9	$15

[a]Average daily "on" hours for wall switch is 14.7. Average daily occupied hours for the office is 12.9.

OPERATING COST COMPARISON
OPEN OFFICE AREA, 1000 SQ. FT.

Performance	*Base Case*	*Time Scheduling*	*Occupancy Sensors*	*Daylighting*	*Time Scheduling + Daylighting*
Annual Energy Use[a]	5700 kWh	5100 kWh	5000 kWh	4200 kWh	3700 kWh
Annual Energy Cost	$340	$305	$300	$250	$220
Annual Energy Cost Savings	—	$35	$40	$90	$120

[a]Average daily "on" hours for wall switch is 9.1. Average daily occupied hours for the office is 6.8.

Cost-Effectiveness Assumptions

Each of the two operating cost comparisons assumes that the workspace has approximately 1.5 watts per square foot of ceiling lighting, with parabolic troffer luminaires containing T-8 lamps and electronic ballasts. Daylighting examples assume a design light level of 55 footcandles at work surfaces. Assumed electricity price: $0.06/kWh, the federal average electricity price (including demand charges) in the U.S.

Figure 2-3. Operating cost comparisons for private office and open office spaces, using various types of controls. *Source: Federal Energy Management Program, U.S. Department of Energy.*

tions for occupancy sensors, such as:

- Ceiling-mounted controls with 360° coverage
- Ceiling-mounted controls with elongated "corridor" coverage
- Wall-mounted controls with a fan-shaped coverage pattern
- Ceiling-mounted controls with a rectangular coverage pattern

Take note of the difference between each device's sensitivity to minor motion (working at a desk) vs. major motion (walking or half-step activity). The manufacturer should provide coverage diagrams for both levels of activity.

Ballast/Control Compatibility

Watch out for mismatched components. For fluorescent lighting, ballasts and controls must be compatible. Fluorescent fixtures which are intended to be dimmed require special dimming ballasts. There are several kinds of control systems, and likewise, and there are several varieties of dimming ballasts.

The two main types of lighting control systems are line voltage and low voltage. Additionally, there are several varieties of low voltage control signals. If you wish to design a fluorescent dimming system, check with the ballast and control manufacturers to ensure that the components will be compatible. Line voltage controls tend to be less expensive, but less flexible, than low voltage controls. If the area does not require low voltage components such as light-level sensors, a line voltage control may be adequate.

Many lighting designers believe that electronic dimming ballasts will be the "future standard" for fluorescent lighting.

Locate Products on Plans

As you locate the controls, ask yourself these questions:

- *Is the placement appropriate?*

Make sure the controls are easy to locate and to access. Don't put them in a closet that might be locked.

Use appropriate controls for each space. Note that the locations of partitions and walls will affect the coverage patterns of sensors. Also pay attention to the locations of doors, air vents, and vibrating machinery. To avoid false triggering, make sure that the sensor coverage will not extend beyond the controlled space. Check for high ceilings. Check that nothing besides the occupants will trigger the sensor, and that the movements of the occupants will always be detected, even if the movements are minute. Seek the manufacturer's assistance if necessary.

- *Are the controls easy to use? Easy to maintain?*

Check that the controlled lighting can be seen from the control panel or switch location. Otherwise, occupants will have to yell to each other "Is that good? Is it dim enough?!" If the controls adapt to the normal behavior of people, they will be accepted. If not, they will be rejected.

Make the control scheme simple. If controls aren't simple, they will not be used. Controls should make sense and provide flexibility to all users.

- *Have you considered security issues?*

In high-security applications, occupancy sensors will indicate that people are present wherever lights are on. It is also advisable that, in these areas, there should be no manual off option and sensors should be protected from tampering and vandalism.

COMPILE CONSTRUCTION DOCUMENTS

A complete set of construction documents includes (but is not necessarily limited to):

- Drawings, showing control locations, circuiting, and a control zone diagram to show which light fixtures are controlled by which device and how the controls are interrelated.

- Wiring diagrams for control components.

- A schedule of controls, showing catalog numbers and descriptions of selected products (including all necessary power packs and accessories).

- Written specifications for the control system, explaining the work and submittals included and clearly describing the approved equipment needed to achieve the desired results.

CONSTRUCTION OBSERVATION

During construction observation, the construction documents are reviewed with the contractor to make sure that the intent of the control system and the method in which it should be installed is understood. The controls manufacturer might provide a training seminar for team members or facilities managers who are not familiar with proper installation and operation of the selected devices.

It pays to make sure the contractor understands the way the control scheme works. In the Way Station project (see Chapter 23), the light-level sensors were supposed to be installed underneath the indirect light fixtures. Instead, they were initially installed on top of them. When the sensors determined that more light was needed, they turned the lights on. But, when the lights came on, they shone on the sensors—so off they went again…

When the installation is complete, the controls are commissioned:

- Light-level or delay-time set points are set
- Dip switches are set
- Sensors are aimed for maximum accuracy
- Preset dimming scenes are set
- The system is tested to make sure it functions as intended

Lastly, users are educated to make sure they know how to use their controls and to get them excited. One great way to familiarize employees with new controls is to provide them with an operator's manual. And, the best way to get the manual right is to invite a group of occupants and facilities managers to contribute to it.

On a final note, watch out for inadequate light levels. Make sure that set points are selected that will please the majority of the occupants. Get their input if possible.

Figure 2-4. Selection of controls for various types of spaces: room by room analysis.

SPACE TYPE	USE PATTERN	IF...	THEN...
Cafeterias or Lunchrooms	Occupied occasionally	Daylighted	Consider daylight-driven dimming or on/off control
		Occupied occasionally	Consider ceiling-mounted occupancy sensor(s). Make sure minor motion will be detected in all desired locations.
Classroom	Usually occupied Occasionally occupied	Multi-tasks like overhead projectors, chalkboard, student note taking and reading, class demonstrations	Consider manual dimming
		Occupied by different students and teachers	Consider ceiling- or wall-mounted occupancy sensor(s) and manual dimming. Make sure that minor motion will be detected.
		Lights left on after hours	Consider centralized controls and/or occupancy sensors.
Computer Room	Usually unoccupied	Lights are left on all the time	Consider occupancy sensors with manual dimming. Be sure that minor motion will be detected and that equipment vibration will not falsely trigger the sensor.
Conference Room	Occupied occasionally	Multi-tasks from video-conferencing to presentations	Consider manual dimming (possibly preset scene control)
		Small conference room	Consider a wall box occupancy sensor
			Consider ceiling- or wall-mounted occupancy

(*Continued*)

Figure 2-4 (Continued)

SPACE TYPE	USE PATTERN	IF...	THEN...
		Large conference room	sensor(s). Be sure that minor motion will be detected in all desired locations.
Gymnasium or Fitness	Usually occupied	Requires varied lighting levels for activities	Consider manual dimming and occupancy sensors. Be sure that the HVAC system will not falsely trigger the sensor.
Hallways	Any	Occasionally or usually occupied	Consider occupancy sensors with elongated throw. Be sure that coverage does not extend beyond the desired area.
		Daylighted	Consider daylight on/off control
Health Care/ Examination Rooms	Occasionally occupied	Different lighting needs for examination	Consider manual dimming
		Small areas	Consider a wall box occupancy sensor
Health Care/ Hallways	Usually occupied	Daylighted	Consider automatic daylight-driven dimming
		Requires lower lighting level at night	Consider centralized controls to lower lighting levels at night
Health Care/ Patient Rooms	Usually occupied	Different lighting needs for watching television, reading, sleeping and examination	Consider manual dimming. Occupancy sensors may not be appropriate
Hotel Rooms	Occasionally occupied	Use primarily in the late afternoon through evening for sleeping and relaxing	Consider manual dimming
Laboratories	Usually occupied	Daylighted	Consider automatic daylight-driven dimming in combination with occupancy sensors.

(Continued)

Figure 2-4 (Continued)

SPACE TYPE	USE PATTERN	IF…	THEN…
Laundry Rooms	Occasionally occupied	Requires high light levels, yet lights are usually left on	Consider occupancy sensors
Libraries/ Reading Areas	Usually occupied	Daylight…	Consider automatic daylight-driven dimming
		Lights left on after hours	Consider centralized controls
Libraries/ Stack Areas	Occasionally occupied	Stacks are usually unoccupied	Consider ceiling-mounted sensor(s)
Lobby or Atrium	Usually occupied but no one "owns" the space	Daylighted and lights should always appear on...	Consider automatic daylight-driven dimming
		It isn't a problem if lights go completely off in high daylight	Consider automatic daylight-driven dimming or on/off control
		Lights are left on all night long, even when no one is in the area for long periods	Consider occupancy sensors. Be sure that minor motion will be detected in all desired areas.
Office, Open	Usually occupied	Daylighted...	Consider automatic daylight-driven dimming
		Varied tasks from computer usage to reading	Consider manual dimming
		Lights left on after hours	Consider centralized controls and/or occupancy sensors.
Office, Private	Primarily one person, coming and going	Daylighted…	Consider manual dimming, automatic daylight-driven dimming, or automatic on/off
		Occupants are likely to leave lights on and occupants would be in direct view of a wall box sensor	Consider a wall box occupancy sensor

(*Continued*)

Figure 2-4 (Continued

SPACE TYPE	USE PATTERN	IF...	THEN...
		Occupants are likely to leave lights on and partitions or objects could hide an occupant from the sensor	Consider a ceiling- or wall-mounted occupancy sensor
Photocopying, Sorting, Assembling	Occasionally occupied	Lights are left on when they are not needed	Consider an occupancy sensor. Be sure that machine vibration will not falsely trigger the sensor.
Restaurant	Usually occupied	Daylighted	Consider automatic daylight-driven dimming
		Requires different lighting levels throughout the day	Consider manual dimming (possibly preset scene dimming)
		Requires different lighting levels for cleaning	Consider centralized control
Restroom	Any	Has stalls	Consider a ceiling-mounted ultrasonic occupancy sensor for full coverage.
		Single toilet (no partitions)	Consider a wall switch occupancy sensor
Retail Store	Usually occupied	Daylighted	Consider automatic daylight-driven dimming
		Different lighting needs for retail sales, stocking, cleaning	Consider centralized controls or preset scene dimming control
Warehouse	Aisles are usually unoccupied	Daylighted	Consider daylight-driven dimming or daylight on/off control
		Lights in an aisle can be turned off when the aisle is unoccupied	Consider ceiling-mounted occupancy sensors with elongated throw. Select a sensor that will not detect motion in neighboring aisles, even when shelves are lightly loaded.

Chapter 3

Lighting Control 101*

By Scott Jordan, Square D

Lighting control ranges from simple wall switches to complex dimming systems networked with other systems.

In some industries, lighting accounts for more than 60 percent of a facility's electrical bill and 40 percent of the total energy bill. Add indirect costs, such as increased loads on cooling systems and increased luminaire maintenance, and the total can be even higher. As a means of offsetting high energy costs, many codes and standards, such as California Title 24 and ANSI/ASHRAE/IESNA 90.1, *Energy Efficient Design of New Buildings Except Low-Rise Residential Buildings*, require some type of automatic lighting control system for all new construction and major renovations. Even when not required by Code, designers often include automatic lighting control as a financial benefit for their clients.

Lighting control can range from simple wall switches to complex dimming systems networked with other building systems. Each lighting control system has a unique set of capabilities and price points. It's usually up to you to decide which system will perform best for the building owner.

Because lighting needs vary with the intended use (for example, lighting offices, corridors, cubicles, and training rooms) and characteristics of the area (such as room size and shape, ceiling height, and availability of natural light), most buildings contain more than one type of lighting control system. Mixing the available technologies often results in the most cost-effective approach.

By combining control methods that include manual, scheduled, and occupancy with the on/off and dimming actions they perform, you can design an effective and economical lighting control system. Let's look at each method and action separately and then see how they can work together.

---

*This chapter originally appeared as an article in EC&M; reprinted here with permission.

ON/OFF OPERATION

It may seem simple, but on/off operation is an area where many designers create an unworkable lighting scheme. For example, consider a metal-halide lighting system. *Restrike time,* which refers to the time it takes a lamp to begin giving off light after being turned on, is crucial for this type of system. Once metal-halides are shut off, they take several minutes to begin giving off light again after being turned back on. If all of your lamps are metal-halide and you shut them off at night, you'll wait 15 min. for a reasonable level of light when you turn them on the next day. By adding other types of light, as well as dedicating certain fixtures to an "always on" configuration, you can reduce the effect of the restrike time. In planning the layout of your lighting controls, make it obvious which lights should not be shut off, and pay special attention to exit path lighting.

DIMMING OPERATION

When you plan dimming, consider how long it takes for a lamp to go from its floor dimming level to 80 percent output. The effective "floor" of dimming for fluorescent lamps is 20 percent—you won't see any energy savings below that level. The effective floor of dimming for metal-halide lights is about 50 percent, because you are effectively restriking the lamp below that level. Be careful where you place your sensors and how you aim them. You want the lights to come on whether a person or a lift truck enters the area, but you don't want adjacent traffic to cause the lights to dim up and down all day. When you dim lights based on ambient lighting, a time delay on the dim-down will eliminate nuisance dimming.

MANUAL LIGHTING CONTROLS

Manual lighting controls range from a single switch to a bank of switches and dimmers, that are actuated by toggles, rotary knobs, push buttons, remote control, and other means. Manual controls are the most cost-effective options for small-scale situations. However, as the size of the lighting system grows, manual controls lose their cost-effectiveness.

But they can still be an important part of a larger plan, as evidenced by the effectiveness of task lighting with manual controls.

SCHEDULED LIGHTING CONTROLS

When you have a predictable occupancy pattern, scheduled lighting controls are often your best option. You can add special manual overrides to make this work when the area needs light outside the normal hours. Manual controls typically work in conjunction with the scheduled controls to override them for a preset time. You should always leave an exit path lit, regardless of the occupancy schedule. If you are unsure whether such a configuration is necessary for your lighting situation, refer to the Life Safety Code, NFPA 101, as well as state and local regulations and fire codes.

OCCUPANCY CONTROLS

The most important thing to consider with occupancy controls is the zone concept. Imagine you have the lighting controls tied into your building's access card reader. When Bob cards in on a Sunday afternoon, you don't want the whole facility to light up. Instead, you want the lights leading to, and inside of, his office to turn on. The copy machine near Bob's office and the water fountain will also power up. Suppose he needs to visit another part of the building. Motion sensors can track his progress and light up the area ahead of him. As he passes into the next zone, the sensors could turn off the lights behind him or leave them on for a preset time (perhaps an hour). However, you don't want the lights to shut off while Bob is sitting at his desk without moving or while he is working behind a partition and beyond the range of the sensors. Occupancy controls, when applied correctly, improve the usability, security, and efficiency of a building. If applied improperly, however, they force the owner to bypass them or remove them altogether.

GOOD ELECTRICAL DESIGN

Regardless of the system you choose, it's important to remember lighting control systems are electrical switching systems with lighting loads. As with any electrical system, you must observe the same Code rules and design practices relating to overload, short-circuit protection, and grounding. However, misapplications of lighting control devices with limited short-circuit current ratings are common occurrences. These underrated devices may remain in service for many years without incident.

COMBINING CONTROL SCHEMES

Many office, retail, or industrial buildings have been successful in using schedule-based systems as the backbone, supplemented by occupancy sensors and manual switches for smaller offices and special-use areas. The backbone system:

- More easily handles the large amounts of power needed for larger areas.
- Switches HID lamps.
- Ties into building automation systems where desirable.

Schedules can accommodate the large number of people who share open areas, while allowing people to override the system for special circumstances or emergencies. However, the schedule system does not work as well for small areas where the variable work schedule of one person may drive the need for lighting. In those cases, an occupancy sensor or manual switch works well. If you are switching exterior lights, you'll probably need a more robust device than what you are using inside.

BUILDING THE BACKBONE

When you lay out the lighting control system, you are building what most designers refer to as "the backbone system." Planning at this stage is crucial to success. Do the electrical design *before* working out the details of the control scheme. To do this right, you need to address the following key considerations:

- *Electrical switching capability*. Be sure your lighting control system can handle the steady-state current, lamp inrush, ballast harmonics, and available fault currents. You'll often have trade-offs among these factors. For example, a "low-harmonic" ballast will result in a higher inrush—which your system might not be able to handle without significant modification.

- *Mounting location*. The "brains" of lighting control systems should be mounted near the lighting panelboards in the electrical closets. In most cases, however, the owner and the installer typically leave insufficient room for this installation. To paraphrase a rule of carpentry, "Measure twice, install once."
- *Schedules and override*. Changing schedules should be easy. Create a flexible design that allows for different schedules for areas of the building with different needs and alternate schedules for weekends and holidays. Be sure to include overrides by wall switch, telephone, or network interface for unusual circumstances.
- *Sufficient circuits and zones*. To maximize savings, zones must be sufficiently small; you don't need to light up an entire floor to accommodate one person who works late. On the other hand, zones that are too small result in extra circuits and installation expense.

It's easy to see why some lighting-controls projects render mediocre results and why others result in systems that owners show off to visitors. By choosing the right combination of controls, you'll have a system that falls into that second category, and by basing that system on a solid electrical plan, you'll provide a reliable system with a low total cost of ownership.

Jordan is a Power Link marketing manager with Square D, Palatine, Ill.

Chapter 4

How to Select Lighting Controls: Where and Why

By the Federal Energy Management Program, U.S. Department of Energy

Lighting controls can save energy and reduce peak demand in offices and other facilities. Controls save money while providing user convenience and an improved lighting environment. There are several different kinds of controls. The choice of control type should be based on lighting usage patterns and the type of space served.

Typical Lighting Control Applications

Type of Control	Private Office	Open Office - Daylit	Open Office - Interior
Occupancy Sensors	++	++	++
Time Scheduling	+	++	++
Daylight Dimming	++	++	0
Bi-Level Switching	++	+	+
Demand Lighting	+	++	++

++ = good savings potential
+ = some savings potential
0 = not applicable

Figure 4-1. Typical lighting control applications. *Source: Federal Energy Management Program.*

Areas with intermittent occupancy are well-suited to occupancy sensors. In large, open office areas with many occupants, scheduled switching ("time scheduling") is often an effective energy-saving strategy. In daylit offices, properly adjusted daylight sensors with dimming ballasts make sense. Because some workers prefer lower lighting levels, bi-level manual switching is another option. Advanced lighting controls can be used for demand limiting to allow building managers to reduce lighting loads when electricity demand costs are high.

Some types of lighting are not well suited to certain controls. For example, daylight dimming and occupancy sensing are not generally appropriate for high intensity discharge (HID) lighting (which requires a delayed re-start), whereas time scheduling is usually a good match for HIDs.

APPROPRIATE ILLUMINATION LEVELS

Proper illumination levels depend on the type of work being performed, and on occupant preference. Recommended illuminance levels for offices range from 30 to 60 footcandles (10.8 lux), but the quality of the visual environment can have a substantial impact on the "appropriate" amount of illumination. In well-designed office spaces, with light-colored surfaces, appropriate task lighting, and careful placement of lights and furniture to avoid glare and shadows, much lower illuminance levels are acceptable, and usually even preferred.

INSTALLATION AND MAINTENANCE

Proper placement and orientation of both daylight and occupancy sensors is essential. Placement of controls should take into account furniture placement as much as possible. Occupancy sensors must be able to sense all occupants to avoid turning off lights while the space is occupied. At the same time, "false-on" incidents can be triggered by an automatic on/off sensor that is exposed to passersby in an adjoining hallway. Daylight sensors that are placed where they are exposed to an amount of daylight not proportionate to the daylight at the desktops being served will not properly control lighting levels (and will likely result in dissatisfied users who may attempt to disable the control system).

Set time scheduling controls so that the switching times and intervals make sense for the occupants and usage pattern of the space. Occupants need to know how to override the schedule easily when needed.

Choose daylight sensors that can be calibrated quickly and easily, and take the time to calibrate them correctly. The dimming adjustment should be easily accessible to the installer and provide an acceptable range of dimming.

Commissioning and calibration of lighting controls are essential if energy savings are to be achieved and maintained. Occupancy sensors with sensitivity set too high can fail to save energy, but occupancy sensors with too low a sensitivity or too short a delay time can be annoying to occupants. Similarly, improperly adjusted daylighting controls can dim the lights too low, causing occupants to override them (e.g., by taping over the sensor), or can fail to dim the lights at all.

Chapter 5

Identifying, Selecting and Evaluating Control Options

By the National Electrical Manufacturers Association, Lighting Controls Council

There is no one right way of determining which lighting control methods are best for a given building. Many approaches have been used successfully, following the format preferred by the using organization or the person specifying the system.

A budget-based method is relatively common. It works well only when all appropriate costs are considered—e.g., while Control X may have a lower initial cost, will it be appropriate if it saves far less energy than Control Y?

Budget-based methods have become popular because so many issues can be reduced to dollar terms. For example, one can begin by examining the issue of manual control versus automatic control. Assuming manual controls will be less expensive initially, will they be able to save as much as automatic controls over the life of the building or for the period of time the building will be retained? Will automatic control be of value when selling it?

Assuming the automatic control is deemed advantageous, the next step would be to determine if on/off controls will be more suitable than output devices. Once again, the full range of issues should be considered in order to help assure selection of the most effective approach, all things considered. And, no matter which decision is made, the next one could relate to centralized systems vs. local systems. If a centralized system is chosen, should it be stand-alone or integrated into others used for building automation or other purposes?

Many potential decisions will be based on assumptions about how people perform and react, what they will and will not remember to do, and so on. As a general rule, do not be optimistic about individuals' ability to take certain action at certain times. In fact, it is precisely because of "human performance" issues that automatic controls have gained such popularity.

SET SPECIFIC CRITERIA

Once the general type of control system has been decided upon, the next step is specifying the specific control devices. In order to do so, it is essential to establish criteria for evaluating options. Some of the key criteria to consider are given below.

Cost-effectiveness

Cost-effectiveness is the basic criterion that will yield the maximum return on investment. Cost-effectiveness is established by applying life-cycle costing techniques, including analysis of economic life, discount rate, investment costs, and savings, as discussed below.

Adaptability

Many important concerns should be raised when assessing adaptability—i.e., how well a control system can be adapted to an existing facility. Some of these concerns are: Do the physical requirements of the new system fit into the existing space? If the space is not readily available, can it be made available? Is construction necessary? Can the new control system be interfaced with existing local controls, or will the controls have to be replaced? Would local controls be more appropriate?

Maintainability

Maintainability refers to the ease with which a system is maintained, something determined through evaluation of two principal factors: in-house maintenance support through training programs and manuals, and the availability of professionally trained maintenance persons employed, licensed, or authorized by the manufacturer. Dust, moisture, or oil on switching and control components, lack of spare parts, and improper control calibration cause the most common control problems and are easily prevented. In larger centralized lighting control systems, the central processor, the computer, field devices, and all the other electronic equipment must be maintained. Diagnostic programs for the various computer components should be required as part of the specifications.

Wiring diagrams of the system also are essential during maintenance procedures; someone who understands the diagrams must be available. Do not assume that a given manufacturer can always provide effective maintenance on a timely basis. Check references to determine how satisfied other users of the proposed system are with the maintenance services provided. Key concerns in this regard are completeness of preventive maintenance, responsiveness and capability of outside maintenance, and the availability of a service maintenance agreement.

Reliability

Reliability relates to two issues: how well the system performs and the way in which it performs. System performance can be determined primarily by talking with other users of the system. They can relate how often breakdowns occur and the time required to restore equipment to its pre-failure condition. (Most modern controls are highly reliable.)

Programmability

Programmability is the degree to which the programming capability of a microprocessor-based lighting control system can be modified. At one time, user programmability was considered almost essential, to help assure that a general program could be made specific to a given building and its unique conditions. Today, each supplier has a variety of programs that can be drawn upon for application, making it easier to find one that closely matches needs. Furthermore, many of these programs are written in such a way that they can be modified relatively easily, either by the user or by the manufacturer.

EVALUATE OPTIONS

Decision-makers must determine specifically which products are best in performing a given function within the context on an overall system designed for a given space. Although some devices may perform the same function, their control technologies may be markedly different from one another. While cost naturally must be a criterion, reliability, maintenance requirements, guarantees, and availability of service, among other factors, can be just as important. One also must consider the cost of installation and materials, especially in the case of retrofit applications.

By quantifying the various benefits, on an approximated basis where necessary, the various pros and cons can be looked at in the same terms. While it is one thing to say that a given control application may save $10,000 per year in energy costs at the expense of a "little productivity," it is quite another thing to say that energy savings will amount to $10,000 but the resulting productivity loss may be worth as much as $25,000 or more.

Table 5-1 lists typical questions that can be asked of vendors, depending on the nature of the system under consideration.

CONSIDER FINANCING OPTIONS

Several important financing options generally are available for retrofitting and installing a new lighting control system. An owner is not limited to an outright purchase. Two of the principal options are *shared energy savings (SES) contracts* and *leasing*.

Shared Energy Savings Contracts

Shared energy savings contracts are a popular type of performance contract. Under this arrangement, a third party—usually an energy service company (ESCO)—selects, installs, and owns the lighting control equipment at the owner's facility, but the owner and ESCO split the energy savings that result. The actual energy use after the improvements is usually subtracted from a baseline estimate, and the savings are then adjusted to reflect current energy prices. For example, if the improvements save 15,000 kWh and the current rate of electricity is $0.08 per kWh, the value of energy savings is calculated at $1,200. If the same amount of energy was saved but the cost of energy increased to $0.10 per kWh, the savings would be calculated as $1,500. Sometimes the savings split between the contractor and the owner remains constant for the duration of the contract; sometimes they vary. When they vary, it usually begins with a larger percentage for the ESCO (such as 80-20) to enable them to regain the capital spent on energy efficiency improvements, then is graduated to a more even split.

Under an SES contract, building owners pay their own energy bills and pay the ESCO the agreed-upon percentage of the savings on a monthly basis. Table 5-2 depicts a five-year shared savings arrangement in which the ESCO receives 80 percent for the first year, but which shifts more savings to the building owner in later years.

Table 5-1. Questions to ask lighting control system vendors.

Financial Arrangements	What is the total installed cost of the system and what does it include? Over what time period must it be paid? Will the supplier provide certain services? What are they? Can the equipment be leased or rented? How much is it? What does the cost include (and exclude)? What are the delivery lead times?
Reliability	How many units of the model under consideration are currently installed? When was the first unit of this model installed? What is the term of warranty and what does it cover? Under what circumstances can the warranty be extended? Will the seller warrant against damage to any other purchaser's equipment? What are the general liability limits and how are claims settled?
Maintenance	How much does it cost and what does it cover? What schedules are available? (Most companies offer many schedules, depending on the response time and coverage desired.) Where is the nearest service office? Have they been trained on the piece of equipment being considered? (If this equipment is the only model of that type installed, chances are that a great deal of on-the-job training will be provided for in-house service people.) What spare parts are recommended? How many different modules does the system have? Does the company selling and installing the equipment also manufacture and service it? Is the system configured so that it can be backed up?
References	Who in the area is using this model equipment? How is it being used? Is the application comparable to the one being contemplated? How long has the equipment been in place?
Training	How much training is required? Is it included in the purchase price? How much does extra training cost? Where does training take place? How long will it last? Is applications support generally required? Can the system be expanded or upgraded easily? How much will expansion cost? What is generally involved?

Table 5-2. Example of a shared energy savings agreement.

Year	*Savings*	*Split*	*Payment to ESCO*	*Building owner's cash flow*
1	$30,000	80-20	$24,000	$6,000
2	$25,000	70-30	$24,500	$10,500
3	$32,000	70-30	$22,400	$9,600
4	$28,000	60-40	$16,800	$11,200
5	$30,000	60-40	$18,000	$12,000

The advantages of the SES approach are:

- both owners and the ESCO benefit from savings;
- the ESCO has an incentive to make the facility as energy-efficient as possible;
- the ESCO is responsible for servicing and maintaining the equipment and has overall project responsibility; and
- building owners do not bear financial risk if the equipment fails to perform as expected.

The disadvantages of SES are:

- building owners are not guaranteed savings;
- the owners are not protected if the energy savings do not materialize, because they must pay the energy bill regardless of results; and
- the ESCO may select control equipment that fails to provide the light levels needed for rapid, accurate seeing, or the desirable degree of flexibility.

Leasing

Under a lease agreement, an investor (lessor) completely finances the purchase and installation of lighting control improvements in a facility. The building owner (lessee) makes monthly payments to the lessor for the use of the equipment. The lessee also is responsible for maintaining the equipment. At the end of the lease agreement, the lessee can purchase the equipment at a predetermined residual value, extend the lease, or have the equipment removed. (Lease/purchase agreements are different in that the lessee is considered the owner of the equipment and thus can obtain tax benefits from depreciation.) Typical lease contracts last from five to ten years.

Leasing differs from SES contracts in several significant ways. First, leasing can only apply to equipment that is considered the personal property of the lessor. In other words, the equipment that is leased must be capable of being moved and used elsewhere. For example, modifications to a building's structure to increase energy efficiency (such as improvements to a building's electrical systems) cannot be part of a lease agreement. Lease arrangements also do not include the service agreements that are usually part of SES contracts.

Another significant difference between leasing and SES contracting is that a lease agreement generally has no provision to guarantee the performance of the equipment. The building owner who enters into a lease agreement assumes all of the technical and financial risks of having the equipment.

The lessee can take some steps to reduce the financial and technical risks incurred under a lease agreement. For one, there is provision for the lessee to terminate the contract early if the results are seriously disappointing. The lessee can also obtain independent engineering analysis before installation of the improvements, and can have the equipment inspected by the consulting engineer prior to acceptance. Purchasing a maintenance contract (these cost from 10 to 15 percent of the equipment cost annually) will also help assure that the equipment runs properly, to minimize some of the risk.

The general advantages of leasing are:

- lessees do not make a significant capital outlay to obtain the equipment;
- lessees can maintain their borrowing capacity, because leases are not counted as debts on the company's balance sheet and do not affect creditworthiness;
- lessees can claim tax benefits for the equipment payments (for operating leases) or depreciation (for financing leases), depending on the structure of the agreement;

- lease payments may be offset by savings from the energy efficiency improvements, allowing lessees to keep their cash flow intact; and
- lessees have cash flow predictability.

The general disadvantages of leasing are:

- lessees assume all technical and financial risks associated with the equipment;
- terminating an unsatisfactory lease is costly, and may be disruptive to building operations;
- the life-cycle cost of leased equipment (considering the lease term and the eventual purchase of the equipment) usually is much more than the cost of an outright purchase;
- only energy efficiency improvements that are removable from the facility can be included in a lease agreement;
- the lessee is responsible for servicing and maintaining the equipment.

Utility Rebates and Incentives

Many electric utilities throughout the nation are providing financial incentives to encourage energy conservation and demand reduction. Their goal is to derive more use from existing generating facilities to forestall the need for building new power plants. In essence, it is less costly to "build" a kW through conservation than through construction.

Not surprisingly, controls are one of the principal energy conservation measures being encouraged by electric utilities. As many experienced "hands" will attest, however, this is almost like "carrying coals to Newcastle," since most controls are so cost-effective without rebates. Nonetheless, the concept of a rebate or other financial inducements unquestionably focuses more attention on controls and conservation, and creates another reason for the investment.

CHOOSE AN ECONOMIC ANALYSIS METHOD

Virtually all decisions about lighting ultimately are based on cost. Although one design approach or one type of system may be preferred to another, these preferences can almost always be expressed in terms of cost factors, such as ease of maintenance and reliability.

Major criteria in selecting a system include the budget provided, the initial system cost, the projected life-cycle cost of the system, and the dollar value of benefits. Other important cost-related criteria are the energy efficiency of the lighting system, its overall energy consumption, its relationship to other building systems, and its flexibility for modification or rearrangement. Still other considerations include existing limitations in modernization projects, building code requirements, and owner preferences or biases.

Economic evaluations must take into account a number of specific factors. These include: the design, components, installation labor, and method of payment that contribute to the initial system cost; utility rebates; alternative methods of acquiring the system hardware; operating and maintenance costs, including energy and parts and labor; inflation; interest rates; tax considerations; the economic life of the system; the discount rate; and the value of tangible and intangible benefits of the system.

These economic factors and the system's overall cost-effectiveness can be analyzed by one of four methods: *simple payback, simple return on investment (SROI), internal rate of return (IRR),* or *savings-to-investment ratio (SIR).*

Simple Payback

Simple payback is used to determine how quickly the savings generated by a modification will pay for its cost. It is expressed as:

$$\text{simple payback} = \text{initial cost}/\text{annual savings}$$

If a lighting controller that costs \$1,000 to install saves \$500 per year, its simple payback is 2 years. If it saves \$750 per year, payback occurs in 1.33 years, or 16 months.

Simple payback also can be applied in new construction, for evaluation of alternatives. Accordingly, it may be found that investing an additional \$1,000 in a better control system will create energy savings or other benefits whose value will pay for the additional investment in a relatively short period of time, or whose annualized value exceeds the additional principal, interest, taxes and insurance (PITI) payment associated with the higher first cost.

Simple Return on Investment (SROI)

Simple ROI is the reciprocal of simple payback. It is expressed as:

simple ROI = annual savings/initial cost

Accordingly, a lighting controller that costs $1,000 to install and saves $500 per year will have a simple ROI of:

$500/yr/$1,000 = 0.50/yr = 50 percent

Internal Rate of Return (IRR)

The internal rate of return (IRR) method is more complex than the simple payback or simple return on investment methods. IRR is the interest rate stated as a percent for which the life-cycle savings are just equal to the life-cycle costs. It is calculated using a trial and error process. Selected compound rates of interest are used to discount the cash flows until a rate is found for which the net value of the investment is zero or close to zero. The calculated IRR is compared to the investor's minimum acceptable rate of return to determine if the investment is desirable.

Savings-to-investment Ratio (SIR)

A savings-to-investment ratio (SIR), also known as benefit-cost ratio, compares the present value of savings to be obtained over an investment's economic life to what it costs today to make the investment. It is expressed as the formula:

SIR = present value of future savings/initial cost

If the SIR is equal to 1.0, it means the present value of future savings is equal to the dollars required today to achieve those savings. If the ratio is less than 1.0, it means the investment will not generate as much money as an easy, safe investment would. If the ratio exceeds 1.0, it indicates the investment will yield a return that is better than the easily obtained return.

SIR is a particularly effective means for evaluating the relative merits of alternative systems.

PRIORITIZE OPTIONS

Once several lighting control options have been evaluated, and their interrelationships are known, certain general priorities become known as well. For example, some options should be implemented before others to reduce expense, some options should be performed at the same time to reduce expense, and others should not be exercised now since implementation of a more comprehensive option at a later date will increase overall benefits to be obtained. When general priorities are identified, specific priorities can be evaluated best in terms of the investment they require and the benefits they deliver.

Section III

ISSUES, TRENDS & CODES

Chapter 6

Lighting Controls: Current Use, Major Trends and Future Direction

By Craig DiLouie, Lighting Controls Association

Lighting automation is now becoming the rule rather than the exception, according to a recent market research study funded by The Watt Stopper and conducted by Ducker Research. The study found that lighting automation is being used in a majority of new construction and renovation projects in the office and school markets. Approximately 65 percent of these projects feature lighting automation.

The research was made available as part of the California Energy Commission's Public Interest Energy Research (PIER) Lighting Research Program—a two-year, $5.2 million research and development program that creates new lighting technology and products that can save energy, reduce peak demand, and reduce pollution for the citizens of California.

The study also found that specifiers and users are very interested in the advantages of controls—primarily energy savings and energy code compliance—but seek simple, low-cost solutions.

Four popular control technologies—building automation systems, lighting control panels, occupancy sensors and daylighting systems—are regarded as effective and relatively problem-free. Occupancy sensors and scheduling systems dominate.

Major potential technology advances regarded as most desirable include standard protocols along with plug-and-play solutions and low-cost electronic dimming ballasts.

Standard protocols and low-cost electronic dimming ballasts were identified as technology advances that would have the greatest impact on lighting control application.

"The top trends in terms of importance to specifiers and end-users is the adoption of standard protocols to enable lighting components to talk to each other, as well as integration of lighting automation systems with building management systems," said David Peterson, Director, Strategic Market Development for The Watt Stopper. "A significant upcoming trend is occupant control of lighting via personal dimming."

Figure 6-1. Occupancy sensors. *Source: Leviton*

THE STUDY

The California Energy Commission's Pier Lighting Research Program, in support of Project 5.4: DALI Lighting Control Device Standard Development, identified its first task to be research of the current use of controls, satisfaction with their use, and receptivity to a standard protocol and the benefits of facilitywide dimming.

The goal of Project 5.4 is to accelerate the use of fluorescent dimming in office and school applications, thereby reducing energy consumption and demand. Its objectives are to define customer needs for automatic control, manual overrides, central monitoring and reporting, load shedding, occupancy sensing and daylight controls in commercial office and school applications. The Digital Addressable Lighting Interface (DALI) protocol, enabling digital lighting networks to be constructed in which all components are interoperable and that provide facilitywide dimming, is therefore being studied.

A research effort was formed to address the above questions by talking to specifiers and users of controls. To accelerate the program, The Watt Stopper, a controls manufacturer, offered to share the results of a study conducted by Ducker Research, which addressed many of these questions. That study, funded by The Watt Stopper, was based on telephone interviews of 158 facility managers, electrical engineers and architects.

Figure 6-2. Dimming panel being installed. *Source: HUNT Dimming*

WHAT IS THE PENETRATION OF AUTOMATED LIGHTING CONTROLS?

Respondents indicate that, on average, more than half of all new commercial new construction and retrofit projects finished over the past two years feature automated lighting controls. In new construction projects featuring automated controls, more than 50 percent of the floor area is covered by automated lighting.

"The education market shows the highest adoption rate for automated controls, particularly colleges, universities and other higher education facilities," said Peterson.

The rate of adoption in retrofit applications is somewhat lower across the board. The largest divergence between new construction and retrofit is in the study's "other" category—library, retail, hospital, government, recreational, industrial. According to the study, nearly 80 percent of new construction projects completed by respondents in these applications over the past two years feature automated lighting controls, while less than half of retrofit projects included them. (See Tables 6-1 and 6-2.)

WHAT ARE THE DEMAND DRIVERS?

The top five factors driving the use of automated lighting controls are:

1. Increasing energy savings
2. Complying with owner requests
3. Compliance with state and national energy codes
4. Providing occupant control capability
5. Obtaining utility rebates and incentives

Study respondents also cited "other" as a very important factor, indicating there are potentially many other factors driving the use of automated lighting.

"Energy savings is the primary driver with the owner having ultimate control," said Peterson.

It is interesting to note that the ability to shed lighting in response to a utility request and to monitor lighting energy usage are not consid-

Table 6-1. Projects utilizing automated lighting control in past two years.

	Projects Utilizing Automated Lighting Control in Past Two Years			
	K-12 Educational	*Higher Education*	*Commercial Office*	*Other**
New Construction Percent Penetration	65.0 percent	71.4 percent	61.8 percent	78.7 percent
Retrofit Construction Percent Penetration	53.1 percent	61.9 percent	57.5 percent	42.8 percent

*Other includes library, retail, hospital, government, recreational and industrial.

Table 6-2. Floor area covered by automated controls in projects featuring automated lighting.

	Floor Area Covered by Automated Controls in Projects Featuring Automated Lighting			
	K-12 Educational	*Higher Education*	*Commercial Office*	*Other**
New Construction Percent Floor Area Covered	59.0 percent	57.6 percent	65.4 percent	62.5 percent
Retrofit Construction Percent Floor Area Covered	50.8 percent	45.2 percent	59.2 percent	45.8 percent

*Other includes library, retail, hospital, government, recreational and industrial.

ered very important, nor is daylighting. Also of interest is growing demand for occupant control of lighting, validated in other studies conducted by the Lighting Research Center and the Light Right Consortium.

WHAT METHODS ARE POPULAR?

The study focused on three lighting automation methods: scheduling, occupancy sensors and daylighting systems.

Scheduling technologies include building energy management systems, time clocks and lighting automation panels. Survey respondents indicated that building energy management systems are most often used for scheduling (39 percent), followed closely by time clocks (35 percent) and also lighting automation panels (26 percent).

Building automation systems are traditionally associated with larger buildings of 100,000 sq.ft. and up. In smaller buildings, lighting control panels and time clocks are more likely to be adopted. This is likely due to initial cost and also because electrical contractors prefer standard devices with readily available parts and applications support, no PCs or special programming tools, and simple commissioning.

Occupancy sensors are, according to the study, the most popular automated lighting control solution for all major building types and are adopted by both large and small buildings. Scheduling systems are also somewhat popular, followed by daylighting systems, which are used much less frequently (see Table 6-3).

HOW DO THE TECHNOLOGIES RATE?

Respondents were asked to rate each technology on a scale of 1-5 on how well it met energy savings expectations, and also how problem-free the performance of the various products have been since installation. A score of "1" meant it exceeded expectations; 3 meant it met expectations; and "5" meant it did not meet expectations.

Scheduling ranked the best in regards to meeting expectations and providing reliable performance; daylighting ranked the lowest in both areas. All technologies were rated as effective and relatively problem-free.

Table 6-3. Incidence of use of various lighting automation solutions.

Incidence of Use of Various Lighting Automation Solutions			
	Scheduling	*Occupancy Sensor*	*Daylighting Sensor*
New Construction			
K-12 Education	48.0 percent	65.7 percent	10.5 percent
Higher Education	48.0 percent	75.4 percent	12.7 percent
Commercial Office	54.3 percent	61.7 percent	11.7 percent
Other*	58.0 percent	67.0 percent	20.0 percent
Retrofit Construction			
K-12 Education	35.2 percent	65.2 percent	11.4 percent
Higher Education	39.2 percent	72.3 percent	10.4 percent
Commercial Office	41.5 percent	59.7 percent	7.5 percent
Other*	58.0 percent	67.0 percent	20.0 percent

*Other includes library, retail, hospital, government, recreational and industrial.

Table 6-4. Respondents rank technologies in regards to expectations and reliability.

Technology	Expectations Scoe	Reliability Scoe
Scheduling using building automation system	2.22	2.09
Scheduling using lighting control panels	2.25	2.15
Occupancy sensors	2.56	2.42
Daylighting controls	2.95	2.55

The respondents were questioned about barriers to adoption of these technologies. For building automation systems, the primary barriers include initial cost and end-user lack of experience with the technology. Initial cost is the primary barrier to lighting control panels and daylighting controls. For occupancy sensors, false offs and delays is the largest barrier to use, along with initial cost.

WHAT ARE THE TRENDS IN THE CONTROLS FIELD?

The study identified five trends influencing the controls field and asked respondents to rate each trend on a scale of 1-5, from extremely important (1) to not important (5). These trends are ranked in Table 6-5.

Table 6-5. Respondents rank controls trends in terms of importance.

Trend	Rating
1. Standard protocols for lighting automation systems	2.36
2. Integration of the lighting automation system with the building management system	2.53
3. Increased need for enhanced occupant control of lighting	3.04
4. Increased demand for flexible use of space	3.06
5. Increased use of architectural daylighting design practices	3.73

Respondents ranked standard protocols as the most important trend primarily for three reasons: The systems would work better together, specification would be made easier, and the process would be simplified and made more convenient. Standard protocols provide assurance that components of the lighting control system would work together, and also provide a common set of base functions and commands accessible to the building automation system.

"Most manufacturers have embraced the cause of interoperability as the best way to service the specifier and user," said A.J. Glaser, president of the Lighting Controls Association and HUNT Dimming, a controls manufacturer. "Popular examples include 0-10 VDC and Phase-Control fluorescent dimming technologies. These open industry protocols ensure compatibility among the various lighting manufacturers, which provides additional choice to the specifier at the front end, while providing options to the owner as it maintains the installation into the future."

The second major trend is integration of the lighting automation system with the building automation system. Respondents indicated this was desirable primarily because centralization provides easier operation of both systems; one technician controlling both systems also provides ease of operation; and higher energy savings can be achieved.

Regarding daylighting, respondents did not see this as a major trend and have not changed their practices because of it. Most agreed with the statement, "As architects begin to use more daylighting, it has an impact," speaking in terms of the future noting that this will have an impact when architects begin to adopt it in greater numbers.

Occupant control was identified as a major trend; respondents were also asked another question related to price sensitivity to more sophisticated lighting options. A choice was provided: Given the installed cost for a traditional parabolic system is $2.00 per sq.ft., which of the following three options would they elect to use to improve lighting quality? (See Table 6-6.)

Table 6-6. Respondents indicate their preference for various lighting options.

	Option	Percent
#1	Use a direct/indirect fixture for $2.50/sq.ft. installed	40.3 percent
#2	Integrate occupancy sensors for $3.00/sq.ft. installed	31.3 percent
#3	Integrate occupancy sensors and provide personal dimming control for $3.50/sq.ft. installed	25.4 percent

Option #1 was desirable to respondents primarily because it represented a lower initial cost. Option #2, however, was desirable primarily because it is "cost effective, a good value." Option #3 was desirable primarily because it increased occupant comfort. The implication of the positive response to personal dimming control is that a significant segment of the market would pay a premium of $0.50 per sq.ft. for it.

HOW WILL POTENTIAL TECHNOLOGY ADVANCES BE RECEIVED?

Respondents were read a list of potential advances in controls and asked whether these advances would help facilitate the use and application of control systems. They responded favorably to all, with the strongest interest being in low-cost electronic dimming ballasts, standard protocols and plug-and-play solutions (see Table 6-7—opposite).

Later, when asked to rank these advances (scale of 1-5, from extremely important to not important), standard protocols ranked highest, then low-cost ballasts, then plug-and-pay solutions (see Table 6-8).

Table 6-8. Study respondents rank the importance of advances that would facilitate the use and application of control systems.

1.	Industry standard communication protocols	2.14
2.	Low-cost electronic dimming ballasts	2.23
3.	Plug-and-play solutions	2.41
4.	One-stop solution such as integrated controls with light fixtures	2.93
5.	Addressable and dimmable electronic ballasts	3.05

Standard protocols were regarded as desirable primarily because respondents felt that this would enable simpler, easier operation, while promoting competition among manufacturers to lower costs. The implication here is that there are currently problems with various control systems working together.

Low-cost electronic ballasts were desirable primarily because "cost effectiveness is always important" and because these ballasts are currently too expensive.

Table 6-7. Study respondents indicate what advances would facilitate the use and application of control systems.

	One-Stop Solution Such as Integrated Controls with Light Fixtures	*Plug-and-Play Solutions*	*Low-Cost Electronic Dimming Ballasts*	*Addressable and Dimmable Electronic Ballasts*	*Industry Standard Communication Protocols*
Yes	69.3 percent	77.8 percent	84.3 percent	62.2 percent	78.6 percent
No	29.9 percent	20.6 percent	15.0 percent	21.4 percent	21.4 percent
Unsure	0.8 percent	1.6 percent	0.8 percent	0.0 percent	0.0 percent
Total	100 percent	100 percent	100 percent	100 percent	100 percent

Chapter 7

Study Finds Adoption of Dimming Systems to Be On the Rise

By Craig DiLouie, Lighting Controls Association

Adoption of dimming systems is slowly increasing as lighting industry participants seek benefits of greater flexibility and energy savings, according to a study conducted by ZING Communications, Inc.

The 2004-2005 Dimming Study, co-sponsored by the Lighting Controls Association, explores attitudes in the specification distribution and contractor sales channel by providing and analyzing survey data from architects, lighting designers, engineers, electrical and lighting distributors, and electrical contractors. The 219-page study is based on a survey distributed to 4,317 industry participants with a 6.7 percent response.

The research suggests that the use of dimming systems is steadily increasing, largely due to lighting industry participants specifying and recommending dimming systems to their clients primarily to provide the benefits of flexibility and energy savings in their projects. The research further suggests that dimming is being used in a broader range of spaces and applications, such as personal control and global control that includes integration with other building systems.

Lighting industry participants largely agree that dimming is perceived as a "green" technology, that daylighting/daylight harvesting is becoming more important as an energy-saving strategy, and that today's manufacturers offer "good products and services."

In addition, lighting designers, architects, engineers and electrical contractors generally regard most types of dimming strategies and equipment to generally meet their performance expectations, with low-voltage master controllers and programming, personal dimming control, centralized dimming control and dimming panels scoring highest.

The research further suggests that distributors are motivated to sell dimming systems and believe that dimming equipment generally raises profit on a project. Electrical contractors are highly comfortable installing dimming equipment and believe they make a good profit on projects that feature dimming.

The three most significant barriers to specification and adoption of dimming systems, say respondents, are cost, complexity of design and installation, and variation in dimming performance by manufacturer and ballast type. A majority of market participants anticipate that they would experience higher sales if these barriers were removed. The research suggests that distributors, in particular, anticipate that their sales would at least double.

Figure 7-1. Dimming system. *Source: Leviton*

WHICH MARKET PARTICIPANTS ARE MOST INFLUENTIAL IN SELECTION OF VARIOUS TYPES OF DIMMING EQUIPMENT?

The research suggests that, overall, engineers and, to a somewhat lesser extent, lighting designers, are most influential in selection of most types of dimming products, although there is indication that electrical contractors are highly influential in selection.

Respondents were asked to rate their own level of influence in selection of dimming ballasts, dimming panels, light sensors, occupancy sensors (when used with dimming system), and dimming controls (wallbox dimmers, etc.).

Respondents were also asked to identify the market participant who most often specifies the dimming systems in their building projects, as well as which market participant who most often commissions the dimming system.

Combined, these metrics indicate a relative degree of influence over specification, selection and commissioning of dimming equipment for each of the respondent groups studied.

Lighting design respondents, on average, rate themselves as highly influential (>4.0 weighted average rating) in selection of dimming ballasts (4.3), dimming panels (4.2) and controls (wallbox dimmers, etc.) (4.3). A majority of lighting designer respondents (80 percent) report that they themselves most often specify dimming systems in their projects. More than one-fifth of lighting designer respondents (22 percent) also report that they most often commission the dimming system, although less than one-third (31 percent) report that manufacturer technicians most often commission the system, and about one-fifth (19 percent) say the electrical contractor most often commissions the system.

Architect respondents, on average, rated themselves as highly influential in selection of controls (wallbox dimmers, etc.) only (4.1). Less than one-half (47 percent) report that they themselves most often specify the dimming system, although more than one-fourth (27 percent) report that the engineer most often specifies the system. In addition, 40 percent of architect respondents say they commission the system as well, while about one-fourth (26 percent) report the electrical contractor most often performs this task.

Engineer respondents, on average, rate themselves as highly influential in selection of all equipment types: dimming ballasts (4.6), dimming panels (4.6), light sensors (4.6), occupancy sensors (4.5) and controls (wallbox dimmers, etc.) (4.5). A majority of engineer respondents (93 percent) report that they themselves most often specify the dimming system. In addition, 41 percent of engineer respondents report that they also commission the dimming system, while about one-fourth (26 percent) say the electrical contractor most often performs this task.

Distributor respondents, on average, do not rate themselves as highly influential in selection of any equipment type. Less than one-half (46 percent) report that the engineer most often specifies the dimming system, while about one-fifth (19 percent) say the lighting designer most often specifies the system and less than one-fifth (17 percent) say the architect most often specifies the system. Less than one-half (47 percent) report that the electrical contractor most often commissions the dimming system.

Electrical contractor respondents, on average, rate themselves as highly influential in selection of light sensors (4.0) and controls (wallbox dimmers, etc.) (4.0). Further results may be surprising. Less than one-half (46 percent) say they themselves most often specify the dimming systems in their building projects, and 60 percent say manufacturer technicians most often commission the dimming systems in their projects.

There appears to be disagreement between the three players on the design team (lighting designers, architects, engineers) about who is most influential in product selection and who most often specifies the dimming system. The most critical question is, "Who most often specifies the dimming systems in your building projects?" since it allows an objective view beyond subjective self ratings regarding influence. The engineer is most often cited by all other surveyed market participants as the party that most often specifies the dimming system (100 score), compared to the lighting designer (55), architect (35) and electrical contractor (25). Therefore, it is reasonable to conclude that the research suggests that the engineer is the most important specifier.

There appears to be further disagreement about the importance of the electrical contractor. Respondents representing the design team, in general, do not perceive the electrical contractor as very influential. When asked who most often specifies the dimming systems in their building projects, those who indicated the "electrical contractor" included only 3 percent of lighting designer, 7 percent of architect, 2 percent of engineer, and 13 percent of distributor respondents. However, 46 percent of electrical contractor respondents say they themselves most often specify the dimming systems on their building projects. This seeming disagreement may be explained by the fact that the electrical contractor may engage in substitutions, putting them in a position of choosing the dimming system. It may also indicate that electrical contractors are responsible for specification in a significant number of projects in which there is no other design authority—that is, no architect, lighting designer or engineer involved in the project. The latter proposition, if true, would indicate a much higher degree of overall influence for the electrical contractor than is otherwise suggested by the research.

It's further interesting to note that while electrical contractors are credited with most often commissioning the dimming system by 19 percent of lighting designer, 27 percent of architect, 26 percent of engineer and 47 percent of distributor respondents, 60 percent of electrical contractor respondents report that manufacturer technicians most often commission the dimming system.

WHAT ARE THE LEADING MOTIVATORS FOR MARKET PARTICIPANTS TO SPECIFY OR RECOMMEND DIMMING SYSTEMS?

The research suggests that flexibility, energy savings and client request are the top motivators across the entire lighting sales channel for market participants to specify or recommend dimming systems.

Respondents were asked to rate the importance of various motivators to specify or recommend these systems on a scale of 1 to 5, with 1 being "not important," 3 being "somewhat important" and 5 being "very important." The motivators include, "give occupants personal dimming control," "client requests it," "add value to the design," "energy savings," "obtain utility rebates and incentives," "ability light the space for different uses (flexibility)," "mood setting," and "extend lamp life." Ratings were compiled to yield a single weighted average response for each motivator for each group of respondents. If the motivator received a score of 4.0 or higher, it is considered to be of high importance.

Lighting designer respondents, on average, rate the ability to light the space for different uses (flexibility) (4.5) and mood setting (4.3) to be of highest importance. Lighting designer respondents are the only respondent group to consider mood setting to be of high importance.

Architect respondents, on average, rate the ability to light the space for different uses (flexibility) (4.6), client request (4.5), energy savings (4.4), and giving occupants personal dimming control (4.1) to be of highest importance. This respondent group considers the highest number of motivators to be of high importance. It is the only respon-

Table 7-1. What is your level of influence over selection of each of the following types of dimming products (including manufacturer) on a typical lighting project, on a scale of 1 to 5, with 1 being "not influential," 3 being "somewhat influential," and 5 being "very influential?"

	Lighting designers	*Architects*	*Engineers*	*Distributors*	*Electrical contractors*
Dimming ballasts	4.3	3.1	4.6	3.5	3.4
Dimming panels	4.2	3.2	4.6	2.9	3.5
Light sensors	3.8	3.3	4.6	3.3	4.0
Occupancy sensors (when used with dimming system to trigger on/off or dimming action)	3.7	3.7	4.5	3.4	3.9
Controls (wallbox dimmers, etc.)	4.3	4.1	4.5	3.8	4.0

dent group to regard giving occupants personal dimming control to be of high importance.

Engineer respondents, on average, rate client request (4.2), energy savings (4.0), and the ability to light the space for different uses (flexibility) (4.0) to be of highest importance. This respondent group has similar motivations to architect respondents, although engineer respondents consider giving occupants personal dimming control to be only somewhat important.

Distributor respondents, on average, do not consider any of the motivators to be of high importance. "Energy savings" ranked highest (3.9).

Electrical contractor respondents, on average, consider energy savings (4.0) and client request (4.0) to be of high importance when specifying or recommending dimming systems. This suggests that electrical contractors, when placed in a position of specifying or recommending dimming to clients, on average regard dimming primarily as an energy-saving strategy.

None of the respondent groups rates extending lamp life, obtaining utility rebates and incentives, and adding value to the design to be of high importance.

WHAT IS THE PREVAILING TREND IN ADOPTION OF DIMMING SYSTEMS? WHAT ARE THE MAJOR TRENDS IN USE?

The research suggests that the use of dimming systems in building spaces is slowly increasing. The research further suggests that lighting designers, architects and, to a lesser extent, engineers are bullish on the trend, while distributors and electrical contractors are less bullish in their outlook, possibly due to their being engaged in a broader scope of lighting transactions than lighting designers, architects and engineers.

In addition, the research suggests that there is a perception of dimming as a "green" technology, that manufacturers offer "good products and services," and that daylighting/daylight harvesting is becoming more important.

In a series of questions, various market participants were asked about the penetration of dimming systems. Lighting designer respondents report that they specify dimming systems in an average 79 percent of their building projects; architect respondents report that they

Table 7-2. How important are the following reasons that you specify dimming systems in building spaces, on a scale of 1-5, with 1 being "not important," 3 being "somewhat important," and 5 being "very important?"

	Lighting designers	*Architects*	*Engineers*	*Distributors**	*Electrical contractors**
Give occupants personal dimming control	3.6	4.1	3.1	3.2	3.7
Client requests it	3.9	4.5	4.2	3.6	4.0
Add value to the design	3.9	3.7	3.4	3.3	3.9
Energy savings	3.8	4.4	4.0	3.9	4.0
Obtain utility rebates and incentives	2.4	2.3	2.8	2.6	2.8
Ability to light the space for different uses (flexibility)	4.5	4.6	4.0	3.7	3.9
Mood setting	4.3	3.7	3.3	3.5	3.3
Extend lamp life	3.7	3.9	3.6	3.4	3.3

*Distributors and contractors were asked, "On a scale of 1-5, with 1 being 'not important' and 5 being 'very important,' what is the importance of each of the following factors to your decision to recommend dimming systems to your clients?"

specify dimming systems in an average 77 percent of their projects; and engineer respondents report that they specify dimming systems in an average 49 percent of their projects. (These numbers may sound high.) Distributor respondents report that, on average, 35 percent of their customers purchased dimming systems over the past year. Electrical contractor respondents report that they install dimming systems in an average of 24 percent of their building projects.

Respondents were asked how they would characterize the trend in use of dimming systems in building spaces, given a choice of rapidly increasing, slowly increasing, holding steady, slowly decreasing or rapidly decreasing.

All lighting designer respondents (100 percent) say that the trend is slowly or rapidly increasing. Three-fourth (75 percent) say that it is slowly increasing, and one-fourth (25 percent) say that it is rapidly increasing.

Similarly, all architect respondents (100 percent) say that the trend is slowly or rapidly increasing, with about two-thirds (67 percent) saying it is slowly increasing and one-third (33 percent) saying it is rapidly increasing.

Seventy-nine percent (79 percent) of engineer respondents say that the trend in use of dimming systems in building spaces is slowly or rapidly increasing, while about one-fifth (21 percent) say it is holding steady. About two-thirds (64 percent) say the trend is slowly increasing, while one-sixth (15 percent) say it is rapidly increasing.

Distributors are the only respondent group that sees the trend decreasing. Seven percent (7 percent) of distributor respondents say the trend is slowly decreasing. Less than one-third (30 percent) say it is holding steady. Less than two-thirds (63 percent) say it is slowly or rapidly increasing. About one-half (48 percent) say the trend is slowly increasing, while one-sixth (15 percent) say it's rapidly increasing.

Sixty-nine percent (69 percent) of electrical contractor respondents say the trend in use of dimming systems in building spaces is slowly or rapidly increasing. One-half (50 percent) say it is slowly increasing, while about one-fifth (19 percent) say it is rapidly increasing. Less than one-third (31 percent) say it is holding steady.

To further identify general trends related to dimming, respondents were given a list of statements and asked to what extent they agreed with them on a scale of 1 to 5, with 1 being "don't agree," 3 being "somewhat agree," and 5 being "totally agree." The result was a series

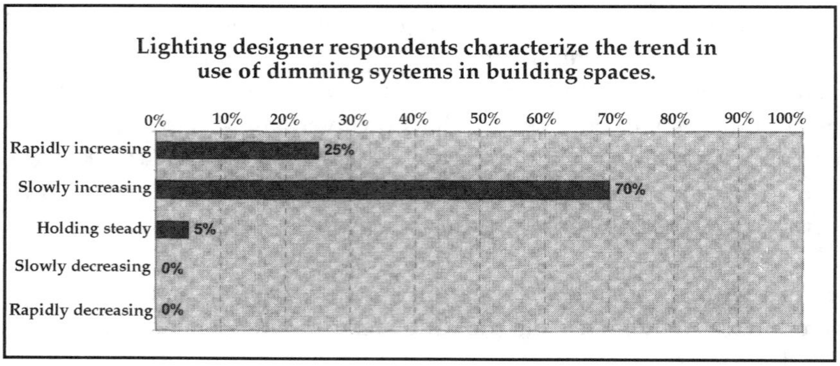

Figure 7-2. Lighting designers characterize the trend in use of dimming systems in building spaces.

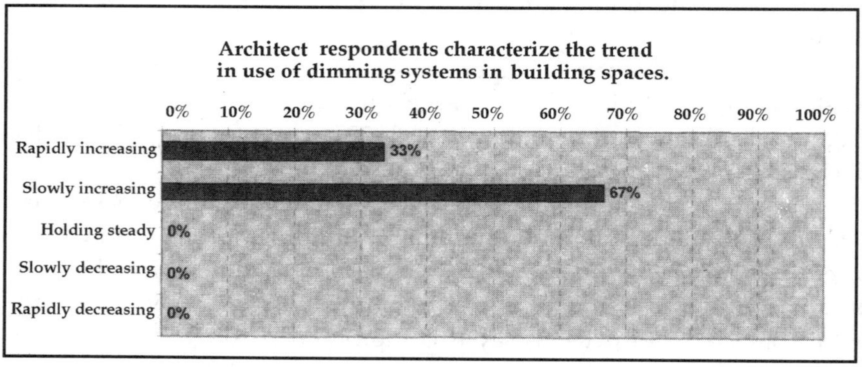

Figure 7-3. Architects characterize the trend in use of dimming systems in building spaces.

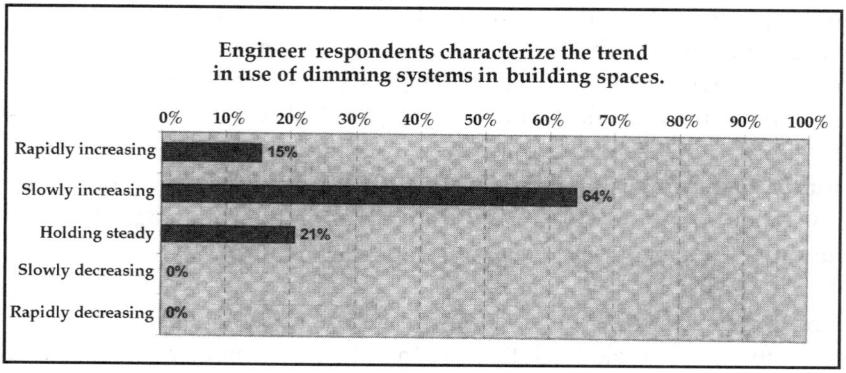

Figure 7-4. Engineers characterize the trend in use of dimming systems in building spaces.

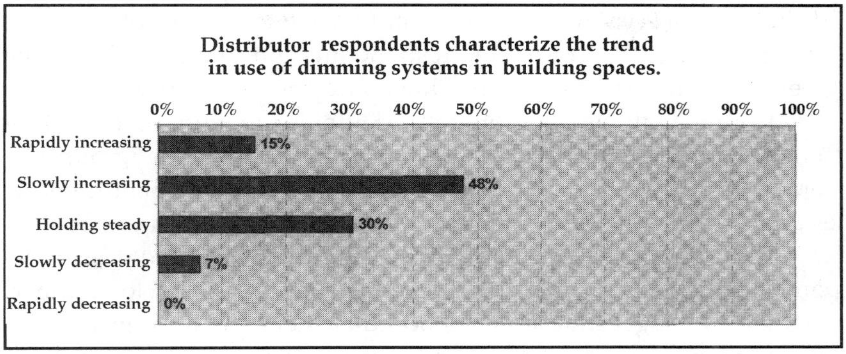

Figure 7-5. Distributors characterize the trend in use of dimming systems in building spaces.

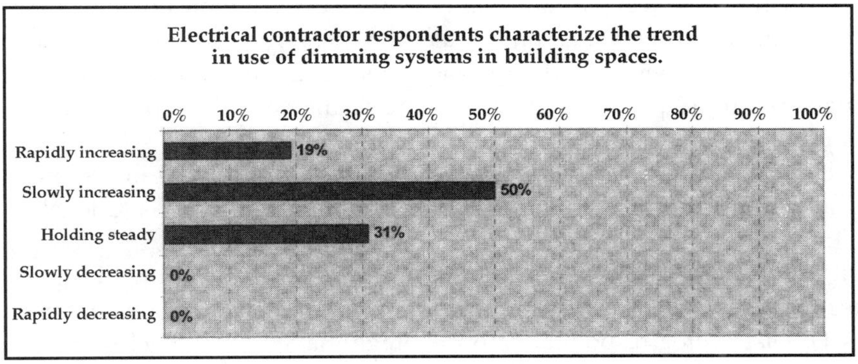

Figure 7-6. Electrical contractors characterize the trend in use of dimming systems in building spaces.

of weighted averages for each statement that are reflective of the average opinion of each respondent group. A rating of 4.0 or higher indicates that the statement, on average, has a high level of agreement by the respondent group.

"Costs are coming down." Lighting designer respondents, on average, have a low agreement with this statement (2.8), while architect and engineer respondents somewhat agree with it (3.0 and 3.2, respectively). Distributors and electrical contractors were asked whether they agree with two statements, whether dimming ballast costs and dimming controls costs are coming down. Distributor respondents, on average, somewhat agree that dimming ballast costs are coming down

(3.1) but have a low level of agreement that dimming controls costs are coming down (2.9). Contractor respondents, on average, somewhat agree that both costs are coming down (3.2 and 3.1, respectively).

"Components are interoperable/Different manufacturers' products work well together as a system." All of the respondent groups, on average, have a low level of agreement with this statement: lighting designer respondents (2.6), architect respondents (2.8), engineer respondents (2.7), distributor respondents (2.0) and electrical contractor respondents (2.8). Of all the statements, this engendered the lowest level of agreement, suggesting a perception that there has been insufficient progress to either make various products successfully interoperable, or educate the market about advances in interoperability.

In a later question, lighting designers, engineers and distributors were asked how well various manufacturer services typically meet their performance expectations on a scale of 1 to 5. The respondent groups indicated that they regard manufacturers' "interoperability with other manufacturers' products" to fall short of meeting their performance expectations (2.4, 2.0 and 2.4, respectively).

"Daylighting/Daylight harvesting is becoming more important." This statement scored in the top three statements in regards to level of agreement. Respondents from the design team perspective, on average, all have a high level of agreement with this statement: lighting designer respondents (4.0), architect respondents (4.6) and engineer respondents (4.1). Distributor respondents, on average, have a low level of agreement with this statement (2.0), while electrical contractor respondents, on average, more than somewhat agree with this statement (3.9).

"Specifiers have enough education to specify dimming systems properly." This statement earned the second lowest level of agreement among all respondent groups. Engineer and distributor respondents somewhat agree with this statement (3.0 and 3.2, respectively), while lighting designer, architect and electrical contractor respondents have a low level of agreement with this statement (2.7, 2.9 and 2.7, respectively). Further, electrical contractor respondents, when asked specifically to what extent they agree with the statement, "Specifiers rarely provide enough or accurate-enough information on drawings," indicated that they more than somewhat agree with the statement (3.6).

"Contractors can install today's dimming systems without difficulty." This statement earned the third lowest level of agreement among respondent groups. The design team, in turn, gives only lukewarm agreement to contractors' ability to install dimming systems without difficulty. Lighting designer and engineer respondents have a low level of agreement with this statement (2.5 and 2.9, respectively), while architect respondents somewhat agree with the statement (3.1). Distributor respondents, on average, similarly have a low level of agreement with the statement (2.7). Electrical contractor respondents, however, on average more than somewhat agree with it (3.7).

In a later question, "lack of skilled labor to install and commission equipment" was presented to respondents as a supposed barrier to specification and adoption of dimming systems, and respondents were asked to rate its importance on a scale of 1 to 5. All respondents except for electrical contractors rated this as somewhat or more than somewhat in importance as a barrier: lighting designer respondents (3.4), architect respondents (3.2), engineer respondents (3.2), distributor respondents (3.0), and electrical contractor respondents (2.9).

"Giving personal dimming control to occupants is a priority for end-users." Architect, distributor and electrical contractor respondents, on average, somewhat agree with this statement (3.2, 3.1 and 3.6, respectively), with electrical contractor respondents, it's interesting to note, having the highest level of agreement. Lighting designer and engineer respondents each have a low level of agreement with this statement (2.9 and 2.8, respectively).

"Dimming is a 'green' technology." This statement earned one of the three highest levels of agreement among the respondent groups. Architect and engineer respondents, on average, have a high level of agreement with this statement (4.2 and 4.1, respectively), while lighting designer and electrical contractor respondents, on average, more than somewhat agree (3.9 and 3.6, respectively). Distributor respondents, on average, somewhat agree with it (3.0).

"Energy savings are fairly predictable with dimming systems." Distributor respondents, on average, more than somewhat agree with this statement (3.7), while lighting designer and electrical contractor respondents somewhat agree with it (3.1 and 3.3, respectively). Architect and engineer respondents have a low level of agreement with the statement (2.9 for each group).

"Dimming systems are reliable." While no respondent group has a high level of agreement with this statement, it scored fourth in level of agreement among respondent groups. Lighting designer, engineer and electrical contractor respondents, on average, more than somewhat agree with this statement (3.9, 3.8 and 3.8, respectively), while architect and distributor respondents somewhat agree with it (3.3 and 3.2, respectively).

"Manufacturers offer good products and service." This statement earned one of the top three highest levels of agreement among respondent groups. Lighting designer respondents, on average, have a high level of agreement with this statement (4.0). All other groups more than somewhat agree with it: architect respondents (3.6), engineer respondents (3.8), distributor respondents (3.8) and electrical contractor respondents (3.5).

"Manufacturer sales reps are knowledgeable and responsive." This statement did not receive as enthusiastic agreement as that given to the manufacturers' products and service. All respondent groups, however, more than somewhat agree with this statement: lighting designer respondents (3.3), architect respondents (3.6), engineer respondents (3.7), distributor respondents (3.3), and electrical contractor respondents (3.6).

"Distributors have all the education they need to sell dimming systems effectively." This statement was posed only to distributors; respondents, on average, more than somewhat agree with this statement (3.8).

"My company earns a good profit when it sells/installs dimming systems." This statement was posed only to distributors and electrical contractors. Both more than somewhat agree with this statement: distributor respondents (3.4) and electrical contractor respondents (3.5).

WHAT MARKETS, LAMP TYPES AND TYPES OF EQUIPMENT ARE COMMANDING THE MOST SPECIFICATION DOLLARS?

The research suggests that lighting designer and architect specification volume is devoted primarily to commercial spaces such as offices, retail, etc., while engineer specification volume is devoted primarily to institutional spaces such as government, schools, hospitals, etc.

The research also suggests that lighting designers and architects are seeing specification dollars most often dedicated to dimming of incandescent lamps, while engineers are seeing specification dollars most often dedicated to dimming of fluorescent lamps.

Table 7-3. Please review the below statements related to dimming systems specified for building spaces, and indicate how much you agree or disagree with the statement on a scale of 1-5, with 1 being "don't agree," 3 being "somewhat agree," and 5 being "totally agree."

	Lighting designers	*Architects*	*Engineers*	*Distributors*	*Electrical contractors*
Costs are coming down	2.8	3.0	3.2	na	na
Dimming ballast costs are coming down	na	na	na	3.1	3.2
Dimming controls costs are coming down	na	na	na	2.9	3.1
Components are interoperable/Different manufacturers' products work well together as a system	2.6	2.8	2.7	2.0	2.8
Daylighting/Daylight harvesting is becoming more important	4.0	4.6	4.1	2.0	3.9
Specifiers have enough education to specify dimming systems properly	2.7	2.9	3.0	3.2	2.7
Contractors can install today's dimming systems without difficulty	2.5	3.1	2.9	2.7	3.7
Giving personal dimming control to occupants is a priority for end-users	2.9	3.2	2.8	3.1	3.6
Dimming is a "green" technology	3.9	4.2	4.1	3.0	3.6
Energy savings are fairly predictable with dimming systems	3.1	2.9	2.9	3.7	3.3
Dimming systems are reliable	3.9	3.3	3.8	3.2	3.8
Manufacturers offer good products and service	4.0	3.6	3.8	3.8	3.5
Manufacturer sales reps are knowledgeable and responsive	3.3	3.6	3.7	3.3	3.6
Distributors have all the education they need to sell dimming systems effectively	na	na	na	3.8	na
My company earns a good profit when it sells/installs dimming systems	na	na	na	3.4	3.5
Specifiers rarely provide enough or accurate-enough information on drawings	na	na	na	na	3.6

In addition, the research suggests that lighting designers and architects most often specify dimming systems for localized applications such as training rooms in their projects, while engineers most often do not specify dimming systems at all.

The research further suggests that lighting designers and engineers most often specify preset-type controls for dimming systems that they specify, while architects most often specify non-preset-type controls.

Finally, the research suggests that lighting designers, architects and engineers most often specify dimming systems that are not integrated with other types of building systems such as occupancy sensors, HVAC, security/proximity, telephone/communications, and PC/networks.

Respondents in the lighting designer, architect and engineer groups were asked to indicate the percentage of their specifications by building space type, lamp type, localized vs. facilitywide systems vs. no dimming system, preset vs. non-preset type, and systems that are integrated into other building systems vs. those that are not.

Forty-four percent (44 percent) of lighting designer respondents report that, overall, their specification dollars are dedicated to commercial spaces such as offices, retail, etc. The remainder is devoted to residential (single-home, multi-family) (30 percent), institutional (government, schools, hospitals, etc.) (22 percent) and industrial (manufacturing, warehouses, etc.) (4 percent). Regarding lamp type, lighting designer respondents report that their specification dollars, overall, are dedicated to incandescent (57 percent), fluorescent (35 percent), HID (3 percent) and other (5 percent). Specification dollars are most often dedicated to dimming systems for localized applications such as training rooms (47 percent), followed by facilitywide dimming systems (lighting control integrated with other types of building control systems) (32 percent). Lighting designer respondents report that, overall, they do not specify dimming systems in about one-fifth (21 percent) of their building projects.

In addition, for those projects where dimming systems are specified, lighting designer respondents report, on average, that they specify preset-type controls in 70 percent of their dimming specifications. Lighting designer respondents, on average, integrate the dimming system into other types of building systems such as occupancy sensors, HVAC, security/proximity, telephone/communications, and

Table 7-4. Overall, what percentage of your dimming specification dollars are for each of the following building spaces...? (Numbers must add up to 100 percent.)

	Lighting designers	*Architects*	*Engineers*
% Commercial (offices, retail, etc.)	44%	44%	38%
% Institutional (government, schools, hospitals, etc.)	22%	18%	49%
% Industrial (manufacturing, warehouses, etc.)	4%	3%	7%
% Residential (single-home, multi-family)	30%	35%	6%
Total Respondents	67	15	51

PC/network in 39 percent of the dimming systems that they specify overall.

Forty-four percent (44 percent) of architect respondents report that, overall, their specification dollars are dedicated to commercial spaces such as offices, retail, etc. The remainder is devoted to residential (single-home, multi-family) (35 percent), institutional (government, schools, hospitals, etc.) (18 percent) and industrial (manufacturing, warehouses, etc.) (3 percent). Regarding lamp type, architect respondents report that their specification dollars, overall, are dedicated to incandescent (50 percent), fluorescent (42 percent), HID (4 percent) and other (4 percent). Specification dollars are most often dedicated to dimming systems for localized applications such as training rooms (58 percent), followed by facilitywide dimming systems (lighting control integrated with other types of building control systems) (19 percent). Architect respondents report that, overall, they do not specify dimming systems in less than one-fourth (23 percent) of their building projects.

In addition, for those projects where dimming systems are specified, architect respondents report, on average, that they specify preset-type controls in 45 percent of their dimming specifications. Architect respondents, on average, integrate the dimming system into other types of building systems such as occupancy sensors, HVAC, security/proximity, telephone/communications, and PC/network in

Table 7-5. Overall, what percentage of your dimming specification dollars is for each of the following types of lighting...? (Numbers must add up to 100%.)

	Lighting designers	*Architects*	*Engineers*
% Incandescent	57%	50%	39%
% Fluorescent	35%	42%	54%
% HID	3%	4%	5%
% Other	5%	4%	2%
Total Respondents	67	15	52

28 percent of the dimming systems that they specify overall.

About one-half of engineer respondents (49 percent) report that, overall, their specification dollars are dedicated to institutional spaces such as government, schools, hospitals, etc. The remainder is devoted to commercial (offices, retail, etc.) (38 percent), industrial (manufacturing, warehouses, etc.) (7 percent) and residential (single-home, multi-family) (6 percent). Regarding lamp type, engineer respondents report that their specification dollars, overall, are dedicated to fluorescent (54 percent), incandescent (39 percent), fluorescent (35 percent), HID (5 percent) and other (2 percent). Specification dollars are most often dedicated to dimming systems for localized applications such as training rooms (37 percent), followed by facilitywide dimming systems (lighting control integrated with other types of building control systems) (12 percent). Engineer respondents report that, overall, they do not specify dimming systems in about one-half (51 percent) of their building projects.

In addition, for those projects where dimming systems are specified, engineer respondents report, on average, that they specify preset-type controls in 56 percent of their dimming specifications. Engineer respondents, on average, integrate the dimming system into other types of building systems such as occupancy sensors, HVAC, security/proximity, telephone/communications, and PC/network in 42 percent of the dimming systems that they specify overall.

Table 7-6. Overall, in what percentage of your building projects do you specify...? (Numbers must add up to 100%)

	Lighting designers	*Architects*	*Engineers*
% Dimming systems for localized applications such as training rooms	47%	58%	37%
% Facilitywide dimming systems (lighting control integrated with other types of building control systems)	32%	19%	12%
% No dimming systems	21%	23%	51%
Total Respondents	68	15	52

HOW MOTIVATED ARE DISTRIBUTORS TO SELL DIMMING SYSTEMS? WHAT IS THE CURRENT LEVEL OF PENETRATION OF DIMMING SALES WITH THEIR CUSTOMERS? HOW IS DIMMING EQUIPMENT TYPICALLY ORDERED AND QUOTED?

The research suggests that distributors are fairly motivated to sell dimming systems and that the presence of dimming equipment generally raises the profit margin on a project. However, while a majority of distributors have a lighting specialist on staff, a minority have a controls specialist on staff, the research suggests, and distributors may need more education.

The research further suggests that distributors most often quote materials for a dimming product quotation through manufacturer-supplied bills of material and price.

In addition, the research suggests that distributors typically order dimming ballasts and controls from the manufacturer rather than keep them in stock. Distributors are most often able to satisfy requests with off-the-shelf items versus dimming components that must be customized for special application needs.

Distributor respondents were asked to estimate the percentage of customers who purchased lighting dimming equipment over the past year through their distributorship. Distributor respondents were also asked whether they have a lighting specialist and a controls specialist on staff; how motivated their salespeople are to sell dimming systems; and whether dimming systems generally raise or lower their profit margins on projects. In addition, distributor respondents were asked how dimming products are quoted, what percentage of dimming ballast and control orders typically are from items in stock compared to items that must be ordered from the manufacturer, and what percentage of dimming orders do customers want a dimming system that includes components that must be customized for special application needs versus off-the-shelf items.

Distributor respondents estimate, on average, that 35 percent of their customers have purchased lighting dimming equipment over the past year through their distributorships.

About six out of 10 distributor respondents (61 percent) report that dimming systems generally raise their overall profit margin on a given project. More than one-fourth (28 percent) say dimming systems have no effect, while about one in 10 (11 percent) say dimming systems reduce their overall profit margin.

In an earlier question, when distributor respondents were asked to what extent they agree with the statement, "My company earns a good

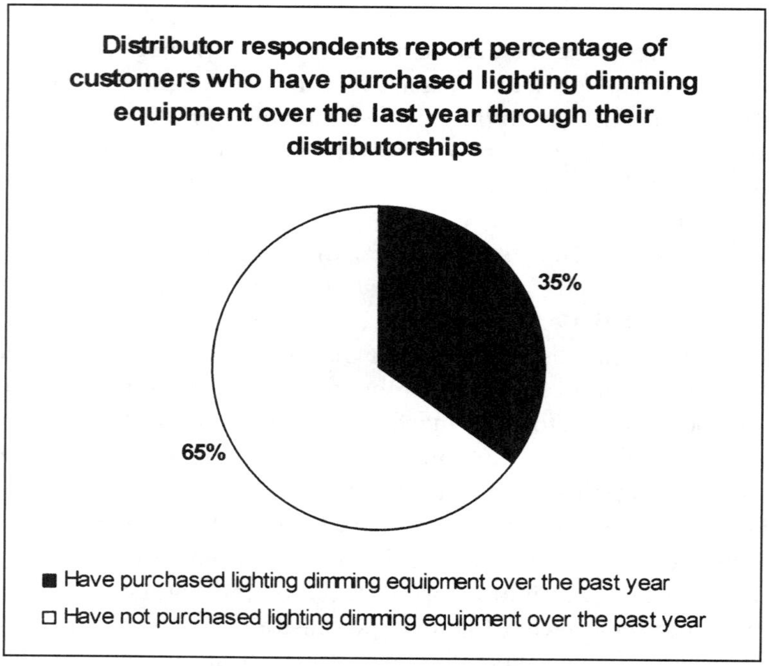

Figure 7-7. Distributors report sales of dimming equipment.

profit when it sells dimming systems," respondents, on average, say they more than somewhat agree with the statement (3.4).

In a later question, however, "insufficient margin on goods sold" was presented to respondents in a list of supposed barriers to adoption of dimming systems; respondents were asked to rate its importance on a scale of 1 to 5. Distributor respondents, on average, regard insufficient margin on goods sold as somewhat important (3.1). This suggests that while distributors may earn a good profit on dimming systems, they would consider a high profit more motivating.

A majority of distributor respondents (86 percent) report that they have a lighting specialist on staff, while forty-one percent (41 percent) say they have a controls specialist on staff. Sixty-eight percent (68 percent) say their salespeople are very or somewhat motivated to sell dimming systems, including dimming ballasts and controls. More than one-half (53 percent) say their salespeople are somewhat motivated to sell dimming systems, while one-sixth (15 percent) say their sales people are very motivated. In contrast, about one-third (32 percent) say their salespeople are not very motivated.

In another question, distributors were presented with a list of statements and asked to what extent they agreed with each statement on a scale of 1-5. In response to the statement, "Distributors have all the education they need to sell dimming systems effectively," respondents, on average, more than somewhat agree with it (3.8).

In a later question, however, "lack of education to properly sell dimming systems" was presented to respondents in a list of supposed barriers to adoption of dimming systems; respondents were asked to rate its importance on a scale of 1-5. Distributor respondents, on average, regard lack of education to properly sell dimming systems as more than somewhat important (3.8).

Forty-four percent of distributor respondents (44 percent) report that a manufacturer-supplied bill of material and price is how they quote dimming systems for a majority of dimming product quotations. Less than one-third (31 percent) say the distributor creates the bill of material and requests a price from the manufacturer. About one-fourth (24 percent) say the electrical contractor creates a bill of material and supplies it to the distributor.

On average, distributor respondents report that 83 percent of their dimming ballast sales and 64 percent of their dimming control sales are ordered from the manufacturer vs. items currently in stock.

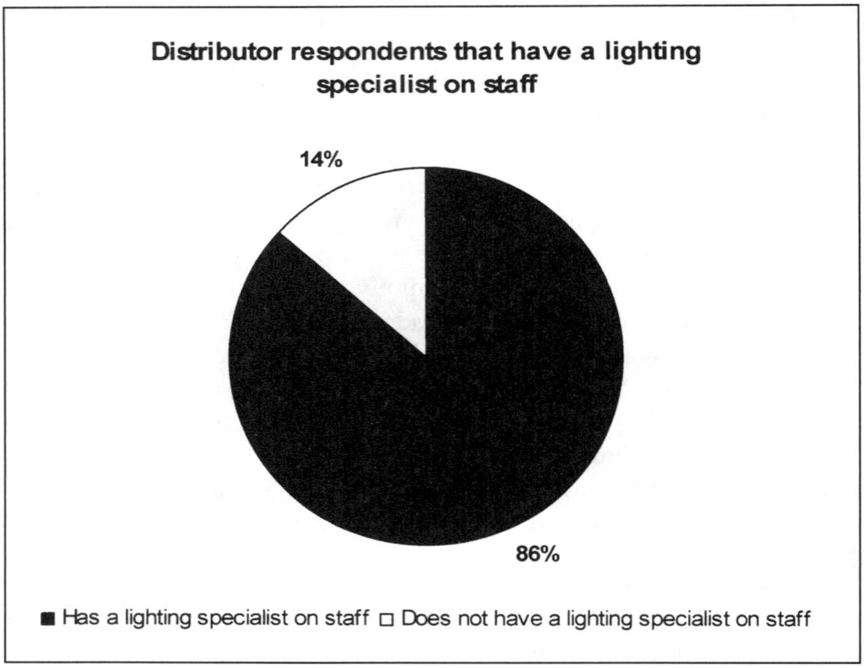

Figure 7-8. Distributors report having a lighting specialist on staff.

On average, distributor respondents report that 66 percent of their dimming orders are off-the-shelf items versus components that must be customized for special application needs.

HOW OFTEN DO ELECTRICAL CONTRACTORS INSTALL DIMMING SYSTEMS, AND HOW OFTEN DO THEY RECEIVE CALLBACKS ON OPERATING PROBLEMS WITH DIMMING SYSTEMS AFTER INSTALLATION?

Electrical contractor respondents, on average, report installing dimming systems in about one-fourth (24 percent) of their building projects.

The research suggests that they earn a good profit when doing so. In an earlier question in the study, when asked to what extent they agree with the statement, "My company earns a good profit when it installs dimming systems," respondents, on average, say they more than somewhat agree with the statement (3.5).

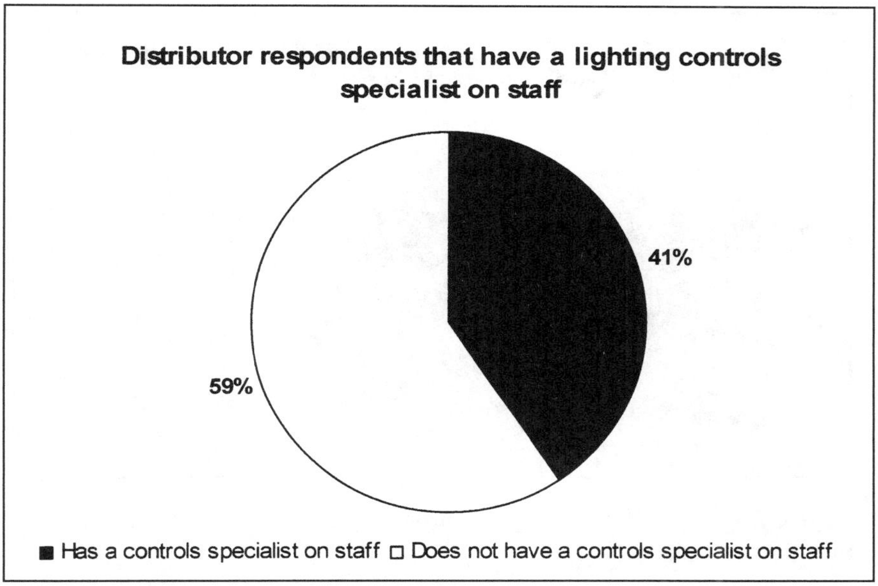

Figure 7-9. Distributors report having a lighting controls specialist on staff.

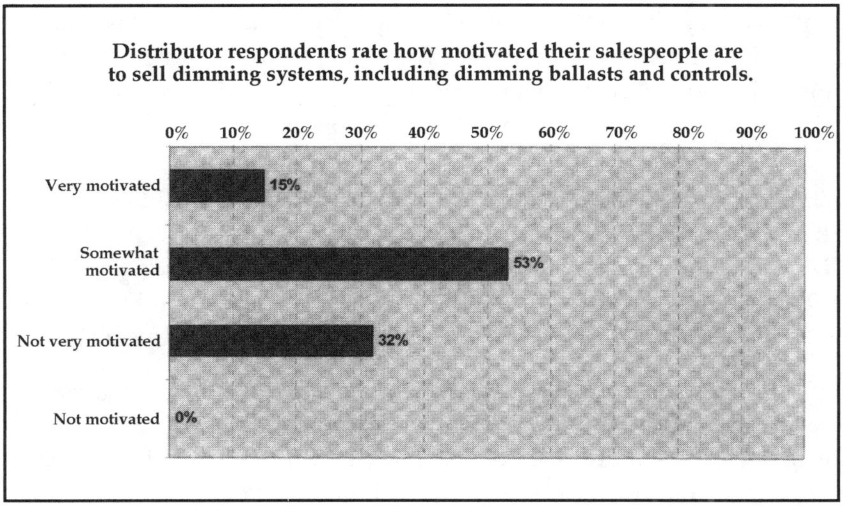

Figure 7-10. Distributors report how motivated their salespeople are to sell dimming equipment.

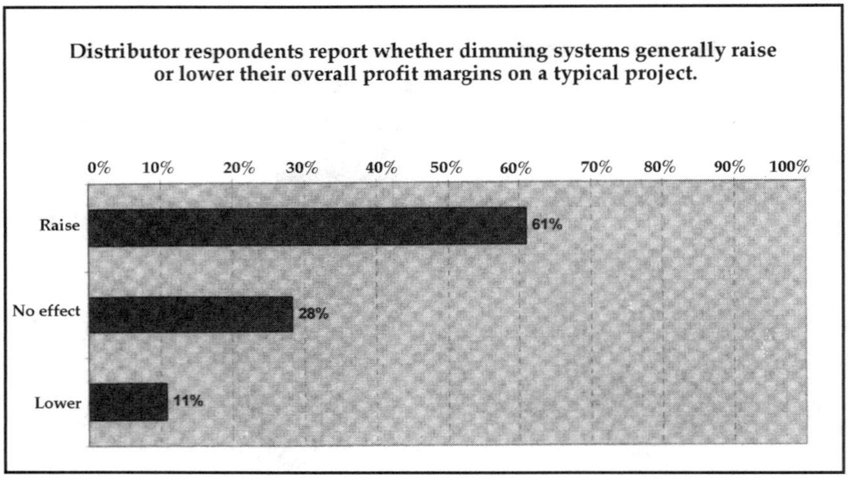

Figure 7-11. Distributors report impact of dimming systems on profit margin.

In addition, electrical contractor respondents, on average, report that they are called back by the customer to fix an operating problem with the dimming system on about one in 10 projects (9 percent). This is slightly higher than the number reported by electrical contractors in another study, the 2004 Commercial Lighting Market Attitudes Study, in which respondents say they are called back to the job site after installation due to lighting system operating problems in an average of 7 percent of their industrial/commercial projects.

WHAT ARE THE MOST IMPORTANT BARRIERS TO SPECIFICATION AND ADOPTION OF DIMMING SYSTEMS, AND WHAT MARKET PARTICIPANTS TYPICALLY PRESENT ROADBLOCKS TO ADOPTION? WHAT WOULD BE THE IMPACT ON SALES IF THE MAJOR BARRIERS AGAINST ADOPTION WERE REMOVED?

The research suggests that the three most significant barriers to specification and adoption of dimming systems are cost, complexity of design and installation, and variation in dimming performance by manufacturer and ballast type. A majority of market participants anticipate that they would experience higher sales if the most important barriers were removed. The research suggests that distributors, in par-

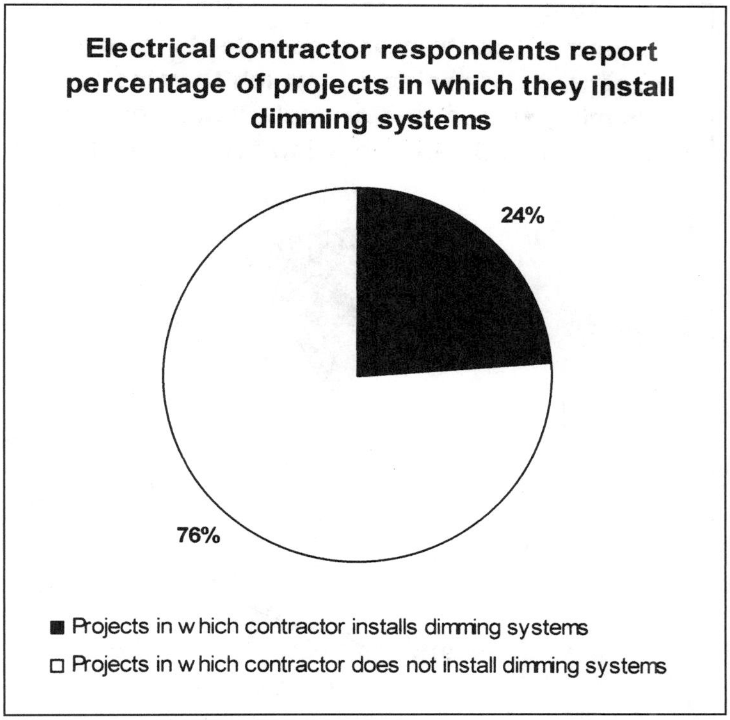

Figure 7-12. Electrical contractors report frequency of projects in which they install dimming systems.

ticular, anticipate that their sales would at least double.

The research further suggests that a significant number of lighting designers, architects and engineers regard the electrical contractor and the owner/client as presenting the most significant roadblocks to the realization of their specification of dimming systems.

Respondents were asked to rate the importance of supposed barriers to specification or adoption of dimming systems on a scale of 1 to 5, with 1 being "not important," 3 being "somewhat important," and 5 being "very important." The result is a weighted average for each barrier by respondent group that is representative of the respondent group. A rating of 4.0 or higher indicates that the barrier is of high importance.

The list of barriers included "initial cost," "complexity of design and installation," "lack of confidence in interoperability of components," "low product reliability," "lack of skilled labor to install and

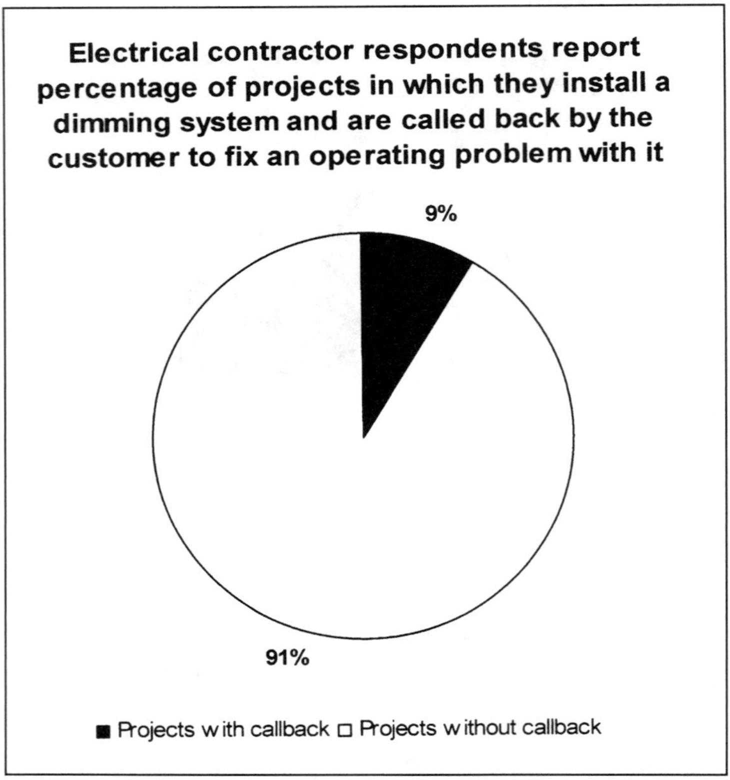

Figure 7-13. Electrical contractors report rate of callbacks for dimming installations.

commission equipment," "long perceived payback period," "lack of customer demand," "commissioning required," "dimming performance may vary by manufacturer and ballast type," "lack of confidence that the system can easily integrate future control technologies," "resistance from other participants in the sales channel," "insufficient margin on goods sold," and "lack of education to properly sell dimming systems."

All respondent groups regard initial cost to be of high importance: lighting designer respondents (4.2), architect respondents (4.0), engineer respondents (4.0), distributor respondents (4.0) and electrical contractor respondents (4.0).

In an earlier question, respondents were given a list of possible trends and statements about dimming, and asked to what extent they agree with each statement on a scale of 1 to 5. When asked to what extent they agree with the statement, "Costs are coming down," lighting designer respondents, on average, have a low agreement (2.8), while architect and engineer respondents somewhat agree with it (3.0 and 3.2, respectively). Distributors and electrical contractors were asked whether they agreed with two statements, whether dimming ballast costs and dimming controls costs are coming down. Distributor respondents, on average, somewhat agree that dimming ballast costs are coming down (3.1) but do not somewhat agree that dimming controls costs are coming down (2.9). Electrical contractor respondents, on average, agree that both costs are coming down (3.2 and 3.1, respectively).

Besides initial cost, only one other barrier is considered to be of high importance, and by only one respondent group. Architect respondents, on average, consider complexity of design and installation to be of high importance as a barrier to specification of dimming systems.

However, nearly all of the respondents, on average, found nearly all the barriers to at least be somewhat or more than somewhat important.

"Low product reliability," "lack of skilled labor to install and commission equipment," and "commissioning required" are the three least important barriers.

Respondents were asked to estimate the impact that would occur on their specifications of dimming systems if the most important barriers were removed.

Three-fourths (75 percent) of lighting designer respondents say they would specify dimming systems more often or much more often if their most important barrier to specification was removed. More than one-half (54 percent) say they would specify dimming systems more often, and about one-fifth (21 percent) say they would specify dimming systems much more often. One-fourth (25 percent) say they would not specify dimming systems more often.

Eight-five percent (85 percent) of architect respondents say they would specify dimming systems more often or much more often if their most important barrier to specification was removed. Sixty-four percent (64 percent) say they would specify dimming systems more often, and about one-fifth (21 percent) say they would specify dimming systems much more often. About one-sixth (15 percent) say they would not specify dimming systems more often.

About three-fourths of engineer respondents (73 percent) say they would specify dimming systems more often or much more often if their most important barrier to specification was removed. About one-half (51 percent) say they would specify dimming systems more often, and more than one-fifth (22 percent) say they would specify dimming systems much more often. More than one-fourth (27 percent) say they would not specify dimming systems more often.

A majority of electrical contractor respondents (80 percent) say they would specify dimming systems more often or much more often if their most important barrier to specification was removed. Forty percent (40 percent) say they would specify dimming systems more often, and forty percent (40 percent) say they would specify dimming systems much more often. One-fifth (20 percent) say they would not specify dimming more often.

Distributors were asked to estimate the impact on sales rather than specification. If the most important barrier were removed, to what extent would their dimming sales increase? Options included same as current sales, double current sales, triple current sales, 4x current sales, 5x current sales and more than 5x current sales.

A majority of distributor respondents (94 percent) believe their sales would at least double if the most important barrier were removed, whereas six percent (6 percent) believe their sales would stay the same. Forty-four percent (44 percent) say their sales would double, about one-third (32 percent) say their sales would triple, one-sixth (15 percent) say their sales would quadruple, and 3 percent say their sales would increase 5x.

Lighting designer, architect and engineer respondents were also shown a list of market participants and asked which typically presents the most roadblocks to their realization of their specification of dimming systems. The list included: lighting designer, engineer, architect, consultant, manufacturer sales rep, building contractor, electrical contractor, distributor, manufacturer, owner/client and "none of the above."

Thirty-eight percent (38 percent) of lighting designer respondents, less than one-half of engineer respondents (46 percent) and about one-fourth of engineer respondents (26 percent) regard the electrical contractor to present the most roadblocks to the realization of their specification of dimming systems. One-fourth of lighting designer respondents (25 percent), 38 percent of architect respondents, and 39 percent of engineer respondents consider the owner/client to present the most roadblocks.

Table 7-7. How important are the following barriers to specifying dimming systems, on a scale of 1-5, with 1 being "not important," 3 being "somewhat important," and 5 being "very important?"

	Lighting designers	*Architects*	*Engineers*	*Distributors**	*Electrical contractors**
Initial cost	4.2	4.0	4.0	4.0	4.0
Complexity of design and installation	3.7	4.1	3.5	3.5	3.4
Lack of confidence in interoperability of components	3.5	3.6	3.5	3.3	3.3
Low product reliability	3.3	2.9	3.4	3.1	3.6
Lack of skilled labor to install and commission equipment	3.4	3.2	3.2	3.0	2.9
Long perceived payback period	3.5	3.7	3.7	2.9	3.1
Lack of customer demand	3.3	2.8	3.6	3.5	3.5
Commissioning required	2.9	3.4	3.3	3.1	3.1
Dimming performance may vary by manufacturer and ballast type	3.7	3.4	3.3	3.3	3.7
Lack of confidence that the system can easily integrate future control technologies	3.5	3.6	3.4	3.2	3.3
Resistance from other participants in the sales channel	3.1	3.4	3.1	3.0	3.0
Insufficient margin on goods sold	na	na	na	3.1	na
Lack of education to properly sell dimming systems	na	na	na	3.8	na

*Distributors and contractors were asked, "How important are the following barriers to adoption of dimming systems, on a scale of 1-5, with 1 being "not important," 3 being "somewhat important," and 5 being "very important?"

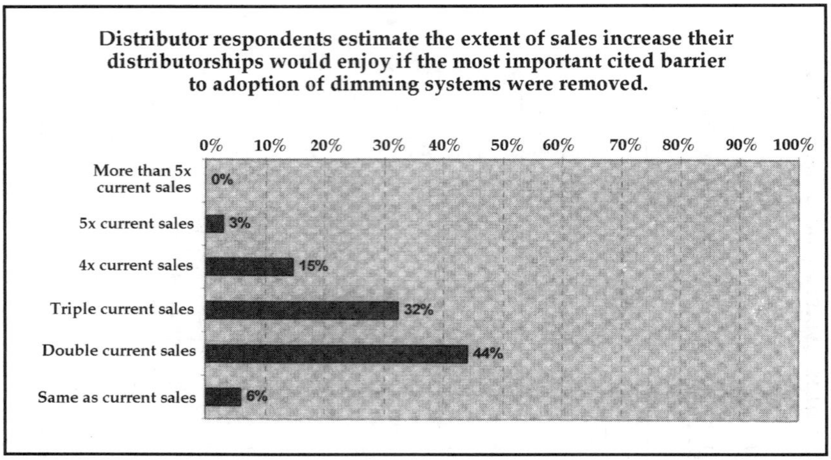

Figure 7-14. Distributors estimate sales potential if major barriers to adoption of dimming systems were removed.

TO WHAT CAUSES DO LIGHTING DESIGNERS, ARCHITECTS AND ENGINEERS ATTRIBUTE ALTERATIONS TO THEIR SPECIFICATION INTENT? HOW OFTEN DO ELECTRICAL CONTRACTORS ENGAGE IN SUBSTITUTIONS OF DIMMING ITEMS, AND FOR WHAT REASONS?

The research suggests that lighting designers, architects and engineers regard budget/cost, delivery/availability and contractor preference for a substituted system to be the most significant reasons the actual installed dimming system may differ from that of the original specification intent.

The research also suggests that electrical contractors believe they do not very often substitute to the original dimming system specifications. When they do, they say it is primarily because of budget/cost and positive experience with the substituted system, presumably due to its being easier to install (higher profit on the job) or demonstrating a high degree of reliability (less likelihood of a callback).

Lighting designer, architect, engineer and electrical contractor respondents were asked, for those occasions that the actual installed dimming system differs from the initial specification intent (design integrity), which three factors are primarily to blame. The list of possible factors includes "budget/cost," "delivery/availability," "specification error," "system compatibility issues," "load compatibility/types," "substituted items are simpler to install and configure by contractor," "contractor had a bad experience with the specified system," "contractor had a positive experience with the substituted system," and "other."

The lighting designer respondents' top three factors are budget/cost (89 percent), positive contractor experience with the substituted system (39 percent), and system compatibility issues (30 percent) and substitutions of items that are easier to install and configure by the contractor (30 percent).

The architect respondents' top three factors are budget/cost (92 percent), delivery/availability (69 percent) and contractor having a positive experience with the substituted system or a negative experience with the specified system (38 percent).

The engineer respondents' top three factors are budget/cost (79 percent), substitution of items that are easier to install and configure by the contractor (42 percent), and delivery/availability (37 percent).

Electrical contractors were shown a list of equipment types and asked how often they substitute against the original specifications for each type on a scale of 1 to 5, with 1 being "not often," 3 being "somewhat often," and 5 being "very often." The result is a weighted average response for each item that is reflective of the attitude of the respondent group. The list of equipment types included dimming ballasts, dimming panels, light sensors, occupancy sensors (when used with dimming systems to trigger on/off or dimming action), and controls.

Electrical contractor respondents, on average, say they do not substitute any of these items very or even somewhat often. They substitute light sensors most frequently (2.8), followed by controls (2.7), occupancy sensors (2.6), dimming panels (2.4) and dimming ballasts (2.3).

Electrical contractors were asked, for those occasions that they substitute items against the original specifications, why they do so, choosing from a list of possible reasons. The reasons include "budget/cost," "delivery/availability," "substituted system did not require programming," "specification error," "system compatibility types," "load compatibility types," "substituted items are simpler to install and configure," "reputation of substituted manufacturer," "bad experience with the specified system," "positive experience with the substituted system," and "other."

Electrical contractor respondents report that budget/cost (38 per-

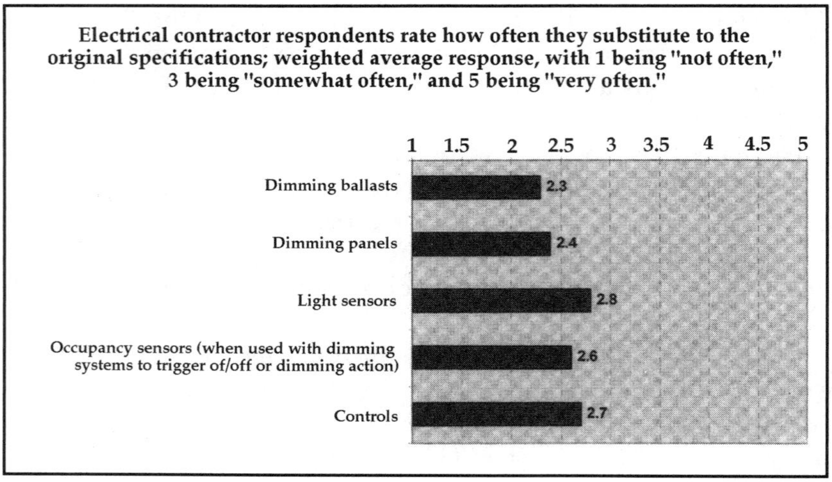

Figure 7-15. Electrical contractors report reasons for substitutions of specified dimming equipment.

cent) and positive experience with the substituted system (38 percent) are the most important reasons they substitute.

HOW DO MARKET PARTICIPANTS RATE VARIOUS DIMMING STRATEGIES AND EQUIPMENT TYPES IN TERMS OF PERFORMANCE?

The research suggests that lighting, designers, architects, engineers and electrical contractors generally regard most types of dimming strategies and equipment to generally meet their performance expectations.

The research further suggests that distributors see some or less than some interest in the market for various dimming strategies and equipment types, based on their sales.

Finally, the research suggests that electrical contractors are more than somewhat or highly comfortable with installing various dimming equipment types.

Respondents were shown a list of dimming strategies and equipment types and asked to rate how well they typically meet the respondent's performance expectations on a scale of 1 to 5, with 1 being "didn't meet expectations," 3 being "met expectations," and 5 being "exceeded expectations." The result is a weighted average response for each strategy or equipment type by respondent group that is reflective of the attitude of the respondent group. A rating of 4.0 or higher marked the result as being of high importance. The strategies and equipment types listed were daylight harvesting, wireless dimming, lumen maintenance dimming, personal dimming control, centralized dimming control, dimming panels, low-voltage master controllers and programming, analog dimming ballasts (0-10VDC, phase control), digital dimming ballasts (DALI, etc.), scheduled dimming, HID bi-level dimming, system integration with other building control systems, and home automation.

No respondent group identified any single dimming strategy or equipment type as highly meeting expectations (4.0 or higher rating).

The four top rated strategies and equipment types across all respondent groups are low-voltage master controllers and programming, personal dimming control, centralized dimming control and dimming panels.

Several ranked below 3.0 and therefore are rated by various respondent groups as failing to fully meet their expectations. These include daylight harvesting (lighting designer and engineer respondents, 2.9), wireless dimming (architect respondents, 2.7 and engineer respondents, 2.6), digital dimming ballasts (lighting designer respondents, 2.9), and HID bi-level dimming (lighting designer respondents, 2.5 and architect respondents, 2.8).

The lowest four ranked strategies and equipment types across all respondent groups are wireless dimming, HID bi-level dimming, daylight harvesting and analog dimming ballasts.

Distributor respondents were asked, when looking at the listed dimming strategies and equipment types, how popular is each, based on their sales, on a scale of 1-5, with 1 being "little interest in the market, 3 being "some interest in the market," and 5 being "hot seller." The result is a series of weighted averages for each strategy or equipment type that is reflective of the group. A rating of 4.0 or higher marked the result as being of high interest in the market.

None of the strategies or equipment types is indicated by distributor respondents as being of particularly high interest in the market.

The top three items in regards to estimated market interest based on distributor respondent sales are personal dimming control, centralized dimming control and dimming panels.

The lowest-ranking three items are analog dimming ballasts, digital dimming ballasts and lumen maintenance dimming.

Electrical contractors were asked how comfortable they are installing the list of dimming strategies and equipment types on a scale of 1 to 5, with 1 being "not comfortable," 3 being "somewhat comfortable," and 5 being "very comfortable." The result is a weighted average for each strategy or equipment type that is reflective of the group. A rating of 4.0 or higher indicates a high level of comfort.

Electrical contractor respondents, on average, report a high level of comfort in installing personal dimming control (4.5), dimming panels (4.3), low-voltage master controllers and programming (4.2), centralized dimming control (4.2), daylight harvesting (4.1), scheduled dimming (4.1) and analog dimming ballasts (4.0).

HOW WELL DO VARIOUS MANUFACTURER SERVICES MEET THE EXPECTATIONS OF MARKET PARTICIPANTS?

Manufacturer services somewhat meet lighting designer expectations, generally do not meet engineer expectations, and somewhat meet distributor expectations, the research suggests.

Lighting designers, engineers and distributors were shown a list of common manufacturer services and asked to rate how well each typically meets their performance expectations on a scale of 1 to 5, with 1 being "doesn't meet expectations," 3 being "meets expectations," and 5 being "exceeds expectations." The result is a weighted average response for each service by respondent group that is reflective of the attitude of the group. A rating of 4.0 or higher indicates that the service has a high level of meeting performance expectations. The list of services includes specification sheets, energy savings projections, educating specifiers/contractors about dimming, educating end-users about the benefits of dimming, marketing support (such as co-op ad dollars, marketing kits, etc.) (distributors only), sales support (manufacturer help to sell big clients and close sales) (distributors only), product availability, equipment delivery, web site, customer service, technical support in the field, technical support 1-800 call-in number, commissioning support, comprehensive offering, and interoperability w/other manufacturers' products.

Table 7-8. How well have the following dimming strategies and equipment types generally met your performance expectations, on a scale of 1-5, with 1 being "didn't meet expectations," 3 being "met expectations," and 5 being "exceeded expectations?"

	Lighting designers	*Architects*	*Engineers*	*Electrical contractors**
Daylight harvesting	2.9	3.1	2.9	3.7
Wireless dimming	3.0	2.7	2.6	3.3
Lumen maintenance dimming	3.0	3.1	3.0	3.7
Personal dimming control	3.3	3.5	3.3	3.8
Centralized dimming control	3.2	3.5	3.3	3.8
Dimming panels	3.4	3.4	3.3	3.7
Low-voltage master controllers and programming	3.4	3.6	3.4	3.5
Analog dimming ballasts (0-10VDC, phase-control)	3.1	3.1	3.1	3.3
Digital dimming ballasts (DALI, etc.)	2.9	3.7	3.3	3.2
Scheduled dimming	3.3	3.5	3.1	3.7
HID bi-level dimming	2.5	2.8	3.0	3.3
System integration with other building control systems	3.0	3.1	3.0	3.8
Home automation	3.2	3.4	3.1	3.3

*Electrical contractors were asked, "How well have the following dimming strategies and equipment types generally met your customers' performance expectations, on a scale of 1-5, with 1 being 'didn't meet expectations' and 5 being 'exceeded expectations'?"

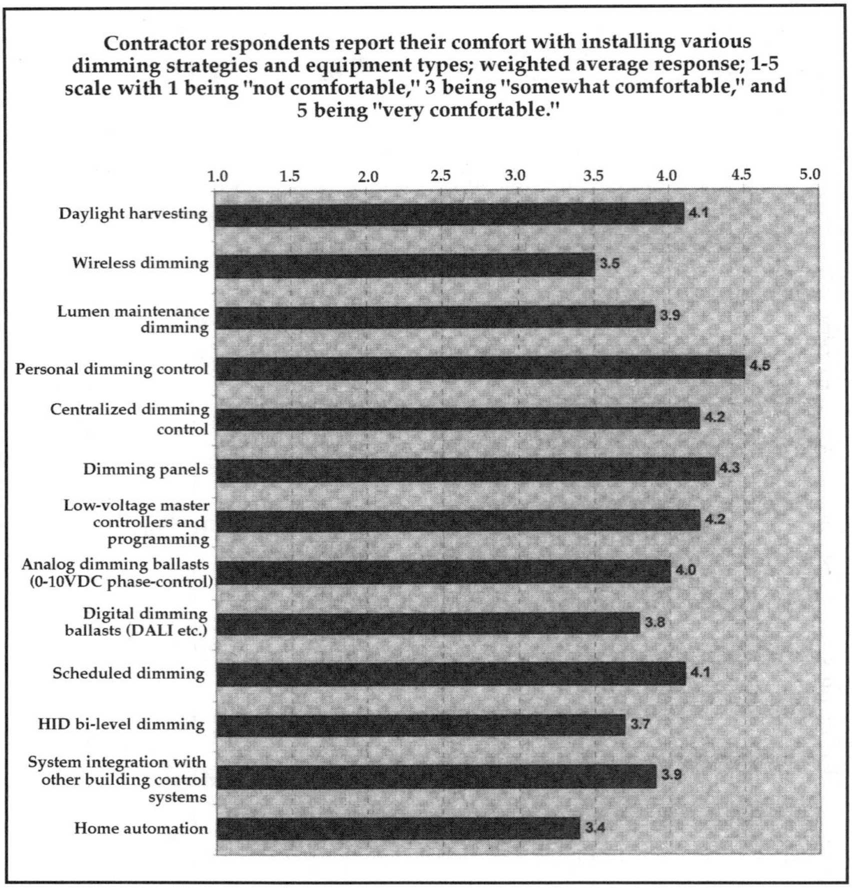

Figure 7-16. Electrical contractors report their comfort level with installing various dimming equipment and implementing strategies.

The respondents, on average, do not consider any of the manufacturer services as particularly exceeding performance expectations.

Lighting designer respondents, on average, regard seven key services as meeting their expectations: specification sheets (3.0), product availability (3.1), equipment delivery (3.1), technical support in the field (3.0), technical support 1-800 call-in number (3.1), commissioning support (3.0) and comprehensive offering (3.1). On average, they regard five key services as falling short of their expectations: energy savings projections (2.4), educating specifiers/contractors about dimming (2.6), educating end-users about the benefits of dimming (2.1), web site (2.9), and interoperability with other manufacturers' products (2.4).

Engineer respondents, on average, regard three key services as meeting their expectations: product availability (3.1), customer service (3.0), and technical support 1-800 call-in number (3.0). On average, they regard 10 key services as falling short of their expectations: specification sheets (2.9), energy savings projections (2.4), educating specifiers/contractors about dimming (2.2), equipment delivery (2.9), web site (2.8), technical support in the field (2.9), commissioning support (2.8), comprehensive offering (2.7) and interoperability with other manufacturers' products (2.0).

Distributor respondents, on average, regard nine key services as meeting or more than meeting their expectations: specification sheets (3.3), product availability (3.1), equipment delivery (3.1), web site (3.2), customer service (3.3), technical support in the field (3.0), technical support 1-800 call-in number (3.7), commissioning support (3.0), and comprehensive offering (3.2). On average, respondents regard six key services as falling short of their expectations: energy savings projections (2.9), educating specifiers/contractors about dimming (2.5), educating end-users about the benefits of dimming (2.5), marketing support (such as co-op ad dollars, marketing kits, etc.) (2.8), sales support (manufacturer helping to sell big clients and close sales) (2.8), interoperability with other manufacturers' products (2.4).

The top three ranked manufacturer services across all three respondent groups are technical support 1-800 call-in number, customer service and product availability.

The bottom three ranked manufacturer services across all three respondent groups are interoperability with other manufacturers' products, educating end-users about the benefits of dimming, and educating specifiers/contractors about dimming.

However, overall, the research suggests that market participants believe that manufacturers offer good products and service. In another question, the statement, "Manufacturers offer good products and service," earned one of the top three highest levels of agreement on a 1-5 scale among respondent groups. Lighting designer respondents, on average, have a high level of agreement with this statement (4.0). All other groups more than somewhat agree with it: architect respondents (3.6), engineer respondents (3.8), distributor respondents (3.8) and electrical contractor respondents (3.5).

Table 7-9. How well do the following dimming product manufacturer-offered services typically meet your performance expectations, on a scale of 1-5, with 1 being "doesn't meet expectations," 3 being "meets expectations," and 5 being "exceeds expectations?"

	Lighting designers	*Engineers*	*Distributors*
Specification sheets	3.0	2.9	3.3
Energy savings projections	2.4	2.4	2.9
Educating specifiers/contractors about dimming	2.6	2.6	2.5
Educating end-users about the benefits of dimming	2.1	2.2	2.5
Marketing support (such as co-op ad dollars, marketing kits, etc.)	na	na	2.8
Sales support (manufacturer help to sell big clients and close sales)	na	na	2.8
Product availability	3.1	3.1	3.1
Equipment delivery	3.1	2.9	3.0
Web site	2.9	2.8	3.2
Customer service	3.2	3.0	3.3
Technical support in the field	3.0	2.9	3.0
Technical support 1-800 call-in number	3.1	3.0	3.7
Commissioning support	3.0	2.8	3.0
Comprehensive offering	3.1	2.7	3.2
Interoperability with other manufacturers' products	2.4	2.0	2.4

Chapter 8

Lighting and LEED

By Craig DiLouie, Lighting Controls Association

Commercial and residential buildings in the United States consume about two-thirds of domestic electric generation, more than one-third of domestic energy production, more than one-tenth of potable water, and 40 percent (or 3 billion tons) of raw materials globally, while producing about one-third of total greenhouse gas emissions and 136 million tons of construction and demolition waste each year.

This model is not sustainable. America's infrastructure depends on an enormous amount of resources, and yet these resources are in finite supply and are steadily diminishing. As a result, a number of leading design firms have committed to sustainable or "green" design practices. The U.S. Green Building Council's Leadership in Energy & Environmental Design (LEED) has become the driving force behind this movement.

LEED defines green design, promotes green design practices, and rewards organizations that adopt green design. LEED projects are certified according to the number of points achieved, indicating how green the building is: Certified (26-32), Silver (33-38), Gold (39-51) and Platinum (52-69).

Lighting is related to achieving at least 8 points and as many as 22 points in these sections: Sustainable Sites, Energy & Atmosphere, Indoor Environmental Quality, and potentially Innovation & Design Process. "Many people don't realize that lighting decisions can actually make a significant impact when working on a LEED project," says Tim Berman, President of Ledalite Architectural Products.

SUSTAINABLE SITES

Sustainable Sites represents 22 percent of the total possible LEED points and intersects with lighting in Credit 8, Light Pollution Reduction (1 point). LEED requires the lighting specifier to "eliminate light trespass from the building and site, improve night sky access and reduce development impact on nocturnal environments."

To gain this point, the lighting specifier must meet or provide lower outdoor light levels than those recommended by IESNA RP-33-99: *Lighting for Exterior Environments*; ensure that the maximum candela value for all indoor lighting must fall within the building (not through the windows); ensure that the maximum candela value for all outdoor lighting must fall within the property; and specify shielding for any fixture within a distance of 2.5 times its mounting height from the property boundary, so that no light spills over the boundary. In addition, all fixtures that produce more than 1,000 initial lumens must be shielded, and all fixtures that produce more than 3,500 initial lumens must meet the Full Cutoff IES classification so no light is emitted skyward.

ENERGY & ATMOSPHERE

Energy & Atmosphere represents 27 percent of the total possible LEED points; lighting plays a significant role in this section. Before earning any points, the specifier must meet two prerequisites. First, all building systems such as lighting control systems must be properly commissioned. Second, the building's electrical systems design must comply with the ASHRAE/IESNA 90.1-1999 model energy code or the local code if more stringent. This is already required in most states. The Department of Energy mandated Standard 90.1-1999 as the minimum design and construction standard for commercial buildings throughout the United States as of July 15, 2004. To date, 32 states have put in place a code at least as stringent as Standard 90.1-1999 (some have adopted stricter codes), while 18 states still have a weaker code or no code at all.

Standard 90.1-1999's lighting requirements are already twice as restrictive as the 1989 standard. For example, the maximum power allowance is 1.3W/sq.ft. for offices, 1.5W/sq.ft. for schools, and 1.9W/sq.ft. for retail buildings. Standard 90.1-1999 also mandates automatic shut-off controls.

In Credit 1, between 1 and 10 LEED points are granted for exceeding Standard 90.1-1999 (or local code) on a scale that rewards maximum energy efficiency. Credit is given based on the whole building's energy use, not just the lighting (see Table 8-1).

Table 8-1. Between 1 and 10 LEED points are granted for exceeding Standard 90.1-1999 (or local code if more stringent) on a scale that rewards maximum efficiency.

New Building	*Existing Building*	*Points*
15%	5%	1
20%	10%	2
25%	15%	3
30%	20%	4
35%	25%	5
40%	30%	6
45%	35%	7
50%	40%	8
55%	45%	9
60%	50%	10

"Lighting as usual is over," says Mark Lien, LC, CLEP, CLMC, Manager-Specification Marketing for Cooper Lighting and a LEED Accredited Professional. "Commodity products will not meet the energy efficiency requirements mandated by LEED-NC in the Energy & Atmosphere category. The credits allowed for exceeding ASHRAE 90.1 require luminaires that have precision optics, use the most efficacious sources, and maximize efficiency."

"The growing popularity of newer technologies that involves such strategies as the dimming of HID fixtures and the use of addressable fluorescent lighting work towards a fully integrated building and support LEED compliance," says Stuart Berjansky, Senior Product Manager, Controllable Lighting for Advance Transformer Company.

"Reducing the energy load is one of the biggest areas for earning LEED points," says Berman. "Using highly efficient fixtures and modern lamp/ballast technology can significant reduce energy requirements. In addition, there are alternative lighting layouts that can be used to lower the average light level in a space while being supplemented with individual task lighting."

Lighting specifiers may also achieve additional points by meeting the requirements under "Credit 3: Additional Commissioning" (1 point) and "Credit 5: Measurement & Verification" (1 point). An independent commissioning authority must review the design and construction documents, commission the systems, and train building operators in system use. The latter requires verification of building performance over time either through site visits or automatic metering.

INDOOR ENVIRONMENTAL QUALITY

Indoor Environmental Quality represents 23 percent of the total possible LEED points. Lighting intersects with this section in two places—controllability of systems and daylighting.

Credits 6.1

Controllability of Systems: Perimeter Spaces (1 point) requires that the design "provide a high level of thermal, ventilation and lighting system control by individual occupants or specific groups in multi-occupant spaces (i.e., classrooms or conference areas) to promote the productivity, comfort and well-being of building occupants." Studies indicate that giving personal control of light levels and thermal comfort to workers can improve their satisfaction. The design should provide at least one lighting control zone per 200 square feet, within 15 feet of the perimeter wall.

Credits 6.2

Controllability of Systems: Non-Perimeter Spaces (1 point) requires the same benefits be provided for occupants in the building's non-perimeter spaces. The design should provide individual lighting controls for at least 50 percent of occupants in regularly occupied non-perimeter spaces.

Credit 8.1

Daylight and Views: Daylight 75 percent of Spaces (1 point) requires that 75 percent of all critical visual task occupied space must achieve a daylight factor of 2 percent, and occupants in 90 percent of regularly occupied spaces must have direct line of sight to vision glazing. Studies indicate that giving occupants access to daylight and access to a view can improve their satisfaction.

"Incorporate daylighting controls when ample daylight is available," says Dorene Maniccia, LC, LEED v.2 AP, Manager, Market Segment Development for The Watt Stopper. "Utilize the expertise of a daylighting or lighting consultant to predict daylight illuminance and distribution, and its effect on lighting quality."

"Daylight harvesting is an increasingly popular strategy," says Berjansky. "It falls into many different areas of the LEED rating system, such as Daylight & Views and Innovation & Design."

INNOVATION & DESIGN

Innovation & Design enables designers with innovative new design approaches to earn from 1 to 4 additional points. "This credit offers opportunities for unique ideas not covered in LEED," says Maniccia. "We've seen occupancy-based plug load controls and DALI control strategies be recognized by the LEED criteria in this category. Because plug loads are exempt from the energy code, and are not addressed by LEED, control strategies that reduce plug loads can significantly help to reduce energy use."

"There are a lot of exciting new technologies in lighting right now," says Berman. "Using advanced technologies can help get credits in this category."

Chapter 9

Lighting and the ASHRAE/IES 90.1-1999 Energy Code

By Craig DiLouie, Lighting Controls Association

Energy codes are designed to set minimum standards for design and construction and can significantly reduce building system life-cycle costs. ASHRAE/IES 90.1 *Energy Standard for Buildings Except Low-Rise Residential Buildings*, developed in the 1970s in response to that era's energy crisis, today is the basis for building codes and the standard for building design and construction throughout the United States; it also influences building designs worldwide.

ASHRAE/IES 90.1-1999, with its tough lighting requirements, became the standard energy code nationwide for all new construction of July 2004.

A provision of the Energy Policy of Act of 1992, put into effect by the U.S. Department of Energy (DOE), required that beginning July 15, 2004, all states must certify that they have energy codes in place that are at least as stringent as Standard 90.1-1999, or justify why they cannot comply. For this reason, 90.1-1999 is written in code language that is readily adoptable by the states.

ASHRAE/IES 90.1-1999 is already the current standard for all Federal building construction, and was adopted for the 2001 version of the International Energy Conservation Code (ICC).

Note that while ASHRAE/IES 90.1-1999 is now the new standard, it sets *minimum* requirements. Individual state energy codes may be tougher and be in compliance with their obligations under the Energy Policy Act.

To see the latest news about code compliance, visit the Building Codes Assistance Project web site: *http://www.bcap-energy.org/newsletter.htm*.

According to the DOE ruling published in *The Federal Register* on July 15, 2002, "Analysis shows, nationally, new building efficiency should improve by about six percent, looking at source energy [where energy is produced], and by about four percent, when considering site energy [where energy is used]."

Four percent load reduction doesn't sound hard overall, but 90.1-1999's lighting requirements are about 50 percent more efficient than the 1989 standard, according to Edward Gray, Director of Energy Policy for the National Electrical Manufacturers Association. In contrast, said Gray, building envelope and HVAC requirements for energy efficiency don't change much.

LIGHTING DIFFERENCES BETWEEN 1989 AND 1999 STANDARDS

Nine out of 10 commercial buildings were constructed before 1986; in most of these older buildings, lighting accounts for 50 percent of electrical energy use, according to the New Building Institute. In newer buildings that meet ASHRAE/IES 90.1-1999, lighting accounts for only 30 percent of electrical energy use.

To address general differences, Standard 90.1-1999 was designed to be easier to use than 90.1-1989 and is written in clearer, mandatory, enforceable language for both new construction and renovations. The code mandates the calculation procedure for fixture wattage to prevent under-calculation, and includes a much broader range of building categories to make the code usable and enforceable. The 1989 code provided single-value whole building lighting power densities for only 11 building types, while 90.1-1999 provides densities for 31 building types. In addition, a number of exemptions in the 1989 version are not present in the 1999 version, such as process facilities; the 1999 version does include a number of narrowly targeted exemptions, such as safety lighting.

Standard 90.1-1999 is largely prescriptive, setting lighting power allowances for interior and exterior applications, with interior applications addressed using either the whole building method or space-by-space method. It provides power limits for exit signs. To address special lighting needs, the code also sets limits for decorative, merchandise, display and accent lighting, and lighting used to reduce glare on computer screen glare in certain spaces. For exterior applications, power allowances are prescribed for building entrances, exits and highlighting. Mandatory tandem wiring requirements are provided to reduce the use of single-lamp ballasts. The lighting power allowances are generally stricter based on advancements in commercially available lighting technologies over the last 10 years.

Regarding the whole building method, for example, office W/sq.ft. is reduced from 2.1-3.3 to 1.3; retail W/sq.ft. is reduced from 2.1-3.3 to 1.9; and school W/sq.ft. is reduced from 1.5-2.4 to 1.5. Regarding the space-by-space method, below are several examples of changes in lighting power allowances:

Lighting Power Allowances (W/Sq.Ft.)

Space	90.1-1989	90.1-1999
Office Enclosed	1.8	1.5
Office Open	1.9	1.3
Conference	1.8	1.5
Training	2.0	1.6
Lobby	1.9	1.8
Lounge/Dining	2.5	1.4
Food Prep	1.4	2.2
Corridor	0.8	0.7
Restroom	0.8	1.0
Active Storage	1.0	1.1

It is assumed that light levels in these spaces will be maintained at IESNA-recommended values, which were used in development of the power allowances in Standard 90.1-1999. Compliance will require more-efficient technology, mostly more-efficient lamps and ballasts. The code provides a table that identifies equipment options (lamps, ballasts, fixtures) with associated percentages of lighting density reductions. For more sophisticated or alternative approaches, engineers can use the energy cost budget method (computer calculations) to demonstrate load reduction within code limits.

ASHRAE/IES 90.1-1999 AND LIGHTING CONTROLS

According to the New Buildings Institute, which developed the *2001 Advanced Lighting Guidelines*, lighting controls can reduce lighting energy consumption by 50 percent in existing buildings and at least 35 percent in new construction.

What Standard 90.1-1999 does is acknowledge that while energy savings vary by application, the positive economic impact of advanced controls is certain. And a broad range of commercially available products and technologies are available from controls manufacturers to address all code requirements and specific opportunities.

Standard 90.1-1999 includes broad mandatory provisions in regards to lighting controls. The 1989 code required minimum controls and covered their accessibility. Automatic controls were addressed in the form of credits for higher power allowances if occupancy sensors, lumen maintenance controls or daylight controls were included in the design.

Facilitywide Lighting Shut-off

Standard 90.1-1999 mandates that either scheduling or occupancy sensing automatic shut-off strategies be used for buildings larger than 5,000 sq.ft., the only exemption being lighting operated 24 hours/day. The control device can be:

- A programmable time scheduling control system for shut off based on time of day when spaces are predictably unoccupied. An independent program schedule is to be provided for areas less than or equal to 25,000 sq.ft., but not more than one program per floor of the building.

- An occupancy sensor that turns the lights off within 30 minutes after the space is vacated.

- An unoccupied/shut-off control signal from another control or alarm system.

Shut-off in Individual Spaces

In addition, each space that is enclosed by ceiling-high partitions must have at least one control device that independently controls the general lighting in the space. Each control device is activated either by an automatic motion sensor or manually by an occupant.

- For spaces equal to or less than 10,000 sq.ft., each control device is limited in coverage area to a maximum of 2,500 sq.ft.

- For spaces greater than 10,000 sq.ft., each control device is limited in coverage area to 10,000 sq.ft.

- Each control device cannot override the time-scheduled automatic shut-off for more than four hours.

- Each control device must be readily accessible and located so that the occupant can see lights from the controlling switch, with an exemption for controls located remotely for safety or security purposes.

Exterior Lighting

Exterior lighting not exempted in the Standard must be controlled by a photocell or astronomical timeclock.

- Other controls required:
- Display/accent lighting
- Display case lighting
- Hotel and motel guest room lighting
- Task lighting
- Non visual lighting (such as for plant growth)
- Demonstration areas

The Watt Stopper provides a helpful illustrative guide in Figure 9-1.

California's Title 24 energy code also mandates bi-level switching to achieve 50 percent energy savings, with exceptions being corridors, storerooms, restrooms, public lobbies, guestrooms, areas with only one fixture, and spaces where occupancy sensors are used.

Building-wide dimming is not addressed by Standard 90.1-1999, although it can be incorporated into computer calculations under the energy cost budget method to demonstrate load reduction.

Automatic Lighting Shut-off Compliance Guidline for ASHRAE 90.1-2001

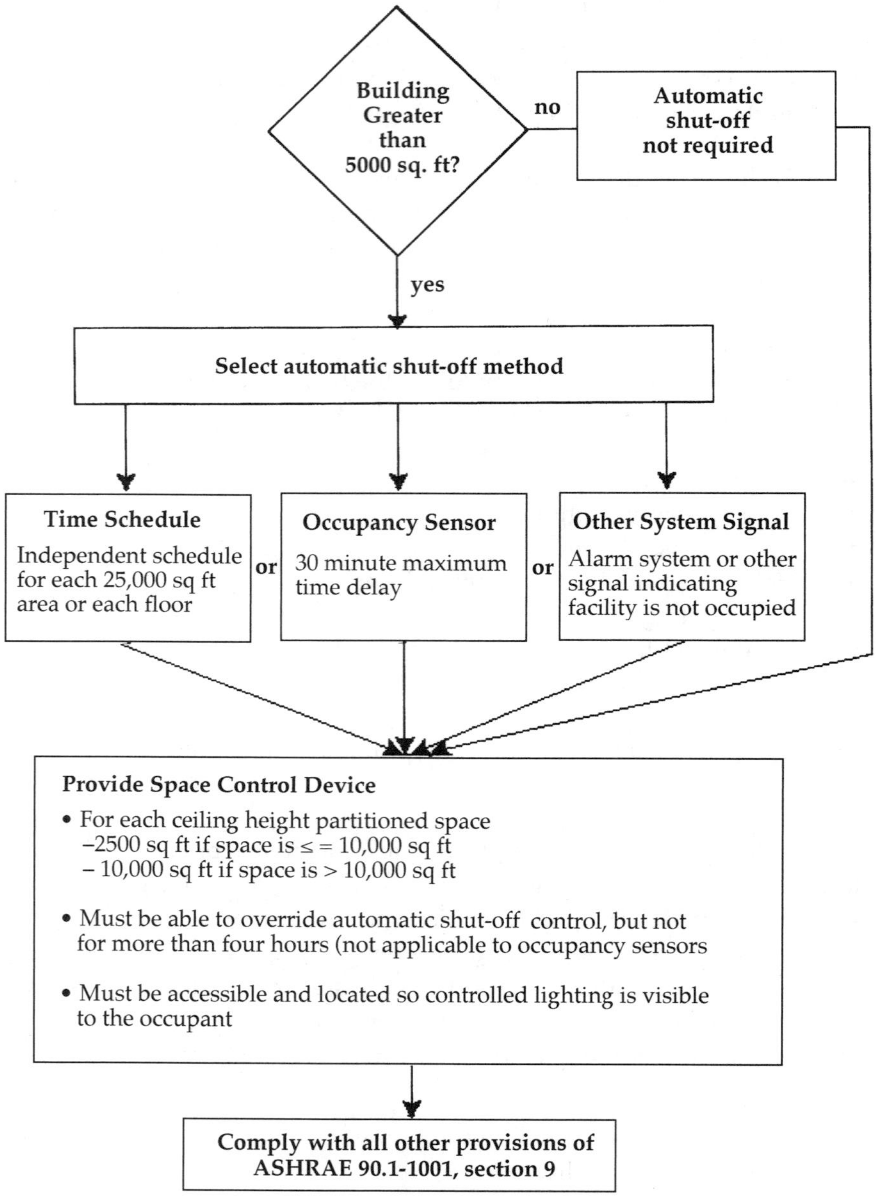

Figure 9-1. Controls and ASHRAE/IES 90.1-1999. *Courtesy of The Watt Stopper.*

Chapter 10

Energy Efficiency Programs Evolve at Utility and State Level

By Craig DiLouie, Lighting Controls Association

Wholesale adoption of maximum energy efficiency measures are estimated to save \$1/sq.ft. (\$0.50/sq.ft. for lighting efficiency)—which translates, based on building statistics compiled by the U.S. Department of Energy, to maximum possible cost savings of about \$60 billion per year while minimizing buildings' impact on the environment.

Unfortunately, the biggest barrier to adoption is the initial cost of the upgrade, including auditing, equipment purchase, installation labor, savings verification, disposal and so on. Energy service companies (ESCOs), which can provide turnkey energy upgrades including financing, tend to focus on the biggest commercial and government buildings as well as schools, leaving smaller organizations struggling to raise the investment capital. Even though these organizations are paying a much higher cost of ownership over the life of their lighting systems, and even though upgrade options are likely available that will produce a lucrative return on investment, the initial cost poses a significant hurdle to investment in efficiency.

Assistance can be found in the form of financial incentives offered by the building owner's utility. According to RealWinWin, Inc., an energy consulting firm based in Philadelphia, more than \$1.5 billion in energy efficiency incentives were available in 2001. That's plenty of money on the table to help lighting professionals and building owners start an energy efficiency project. Incentives escalated in the early 1990s, peaked in 1993, and then declined until renewed interest due to the energy crisis of 1998-1999, which caused an increase in annual funding (see Table 10-1).

In this chapter, we will examine the reasoning behind the continued propagation of utility energy efficiency incentive programs, state efficiency programs that have replaced utility programs in some states, and the evolution of a form of utility incentive called demand response.

Table 10-1. Electric Utility Demand-Side Management Programs, 1989-2000. *Source: U.S. Department of Energy*

Year	*Actual Peakload Reductions[1] (megawatts)*			*Energy Savings (million kilowatt-hours)*	*Costs (thousand dollars[4])*
	Load Management[2]	*Energy Efficiency[3]*	*Total*		
1989	NA	NA	12,463	14,672	872,935
1990	7,911	[5]5,793	13,704	20,458	1,177,457
1991	8,767	[5]6,852	15,619	24,848	1,803,773
1992	7,357	[5]9,847	17,204	35,563	2,348,094
1993	10,583	[5]12,486	23,069	45,294	2,743,533
1994	10,922	[5]14,079	25,001	52,483	2,715,657
1995	13,753	[5]15,807	29,561	57,421	2,421,261
1996	12,965	[5]16,928	29,893	61,842	1,902,197
1997	11,958	13,326	25,284	56,406	1,636,020
1998	13,640	13,591	27,231	49,167	1,420,920
1999	13,003	13,452	26,455	50,563	1,423,644
2000	10,027	12,873	22,901	53,701	1,564,901

[1]The actual reduction in peak load reflects the change in demand for electricity that results from a utility demand-side management program that is in effect at the time that the utility experiences its actual peak load as opposed to the potential installed peakload reduction capability. Differences between actual and potential peak reduction result from changes in weather, economic activity, and other variable conditions.

[2]Load Management includes programs such as Direct Load Control and Interruptible Load Control, and beginning in 1997, "other types" of demand-side management programs. Direct load control refers to program activities that can interrupt consumer load at the time of annual peak load by direct control of the utility system operator by interrupting power supply to individual appliances or equipment on consumer premises. This type of control usually involves residential consumers. Interruptible load refers to program activities that, in accordance with contractual arrangements, can interrupt consumer load at times of seasonal peak load by direct control of the utility system operator or by action of the consumer at the direct request of the system operator. It usually involves commercial and industrial consumers. In some instances, the load reduction may be affected by direct action of the system operator (remote tripping) after notice to the consumer in accordance with contractual provisions. "Other types" are programs that limit or shift peak loads from on-peak to off-peak time periods, such as space heating and water heating storage systems.

[3]Energy efficiency refers to programs that are aimed at reducing the energy used by specific end-use devices and systems, typically without affecting the services provided. These programs reduce overall electricity consumption, often without explicit consideration for the timing of program-induced savings. Such savings are generally achieved by substituting technically more advanced equipment to produce the same level of end-use services (e.g., lighting, heating, motor drive) with less electricity. Examples include high-efficiency appliances, efficient lighting programs, high-efficiency heating, ventilating, and air conditioning systems or control modifications, efficient building design, advanced electric motor drives, and heat recovery systems.

[4]Nominal dollars.

[5]From 1989 to 1996, Energy Efficiency includes "other types" of demand-side management programs. Beginning in 1997, these programs are included under Load Management.

NA=Not available.

Web Page: *http://www.eia.doe.gov/fuelelectric.html.*

Sources: • 1989-1999—Energy Information Administration (EIA), Electric Power Annual, annual reports. • 2000—EIA, Form EIA-861, "Annual Electric Utility Report."

CURRENT STATUS OF ENERGY EFFICIENCY PROGRAMS

Deregulation of the $300 billion electric power industry, begun in 1992, has dramatically changed the landscape for energy efficiency programs. In states that are still regulated, utility energy management programs may be offered by the local utility, in some cases mandated by regulators. In states that are deregulated or in the process of becoming deregulated, the utilities may offer some form of incentive on their own, and the state's deregulation law may pose a "public good" surcharge to raise capital that in turn is reinvested in customer efficiency.

Three types of programs in which financial incentives are offered include public purpose energy efficiency programs, utility energy efficiency programs and utility demand response programs.

Public Purpose Energy Efficiency Programs

These programs are administered either by utilities, state agencies or other third parties and are paid for by utility ratepayers, typically through a non-bypassable System Benefit Charge which is instituted as part of restructuring legislation or rules.

Utility Energy Efficiency Programs

These programs are administered by the local utility and paid for by utility ratepayers through their bundled rates.

Demand Response Programs

These are programs which provide incentives to reduce load in response to system reliability or market conditions, mostly designed for customers with large loads who can make significant reductions at the utility's request. Demand response is favored by many utilities because it addresses their primary problem, which is bridging the gap between wholesale prices and retail prices and ensuring reliability of supply during these periods. Utilities would rather give a customer an incentive to reduce its load during such events rather than pay the customer to reduce its overall load, which can result in lost revenue for the utility. This typical form of demand response incentive is very similar to interruptible rate programs.

Another form of demand response program is based on a real-time pricing contract. After the customer's baseline load is established, the utility charges the customer for this load at a fixed rate. If the customer falls below the baseline, it receives a financial credit. If it rises above the baseline, it must pay market-based pricing. This scenario passes the risk of unstable market conditions on to the customer, who could end up absorbing the cost of wholesale price spikes. To address this risk, customers can acquire price protection products as a hedge. In addition, the utility offers a provision in the agreement for notice of the curtailment event, usually hour-ahead or day-ahead notice.

Depending on the program, energy efficiency programs, which may be applicable to retrofit, new construction and major renovations, can therefore offer:

- Design and engineering assistance, from free energy auditing to savings verification
- $ amount for installing products from a list of approved technologies/products, with a rebate based on the product cost or total installed cost
- $ amount per kWh of energy saved over a year
- $ amount per kW of load removed
- Custom measures, which include new technologies and strategies that are either unproven or in which the energy savings are high variable, which usually must demonstrate through modeling that they save energy and therefore qualify for an incentive
- $ amount credit for reducing load by a set amount when the utility declares a "curtailment event," usually occurring during peak demand periods

LIGHTING REMAINS THE IDEAL RETROFIT

Because lighting is regarded as a relatively "easy" energy efficiency measure with very high energy savings potential, virtually all energy efficiency programs include lighting, with many having specific incentives just for lighting.

Lighting upgrades can affect load in two ways:

- Remove kW from the lighting system via system upgrades including energy-efficient lamps and electronic ballasts, or through creative lighting design strategies

- Reduce kWh by using lighting controls to automatically switch or dim the lights based on occupancy, time of day or in response to utility curtailment needs

Most lighting technologies are proven and qualify for incentives, including products such as occupancy sensors whose energy savings are variable. Facilitywide dimming can be employed to support participation in demand response programs by reducing the lighting load when the utility declares a curtailment event.

Example

A 100,000-sq.ft. office building operating a lighting load of 140kW for 4,000 hours/year upgrades its lighting systems using new T8/electronic ballast lighting systems with state-of-the-art lighting controls. Lighting energy consumption is reduced by 50 percent, or 280,000kWh, saving $28,000 per year at a local average rate of $0.10/kWh. The building is located in the service territory of a utility that offers an incentive of $0.10/kWh for energy efficiency retrofits, which results in the building's owner receiving a financial incentive of $28,000 to perform the upgrade. With a total retrofit cost of about $90,000, this incentive comprises almost one-third of the total initial cost, and dramatically compresses the payback period from about three years to about two years before positive cash flow is realized.

Example

Our 100,000-sq.ft. office building also participates in a demand response program designed to ensure stable market conditions and reliability of power during shortages and other emergencies. At the utility's request, the building must curtail its load to gain the incentives, which can reach $0.35/kWh reduced during shortages. Using a lighting management system that provides facilitywide dimming, the building reduces its lighting load by 25 percent during three summer curtailment events for a total of 15 days, or 5,775kWh. This earns the facility savings of \$577.50 based on the average kWh rate of \$0.10/kWh, and an additional \$2,021.25 for the curtailment events.

A complete lighting upgrade, including new lamps, ballasts and advanced controls, can qualify for all types of energy efficiency incentives, including custom measures. Lighting controls can be used to qualify not only for one-time incentives, but ongoing incentives through certain demand response programs.

Chapter 11

Commercial Lease Properties: Finding the Benefit of Energy-Efficient Lighting Upgrades

By Craig DiLouie, Lighting Controls Association

More than 4.7 million commercial and government buildings, representing over 67 billion sq.ft., currently account for about 25 percent of the nation's energy bill, spending $26 billion annually. A significant number of these buildings and floorspace (23-24 percent) are non-owner-occupied. With an average building age of 30.5 years and average annual energy cost of about $16.4 billion or 1.06/sq.ft., non-owner-occupied buildings are a prime opportunity for upgrade to energy-efficient building technologies—although traditionally, in general, they have been slow to embrace energy efficiency.

Of all building upgrades, lighting is generally considered the easiest and most lucrative. According to the U.S. Department of Energy, technologies developed during the past 10 years can help cut lighting costs by 30 percent to 60 percent while enhancing lighting quality and reducing environmental impact. And according to the New Buildings Institute, which developed the *2001 Advanced Lighting Guidelines*, lighting controls can reduce lighting energy consumption by 50 percent in existing buildings and at least 35 percent in new construction. The energy savings potential of the commercial real estate market, however, remains largely unrealized. The energy-efficient products industry must understand this market to overcome its barriers to capital investment in efficiency, and building owners must understand that there is significant money on the table for them.

LANDLORDS & TENANTS

In a lease property scenario, the owner regards its building as an income-producing asset. Net operating income in turn provides the

Table 11-1. Owner Vs. Non-Owner-Occupied Buildings in the United States. *Source: 1999 Commercial Buildings Energy Consumption Survey, Energy Information Administration (DOE)*

	Buildings (Thousand)	*Floorspace (Million Sq.Ft.)*	*Median Sq.Ft./ Bldg. (Thousand)*	*Median Bldg. Age (Years)*
Nongovernment Owned	4,135	54,994	5.0	30.5
Owner Occupied	2,800	37,785	5.0	29.5
Non-Owner Occupied	1,099	15,596	5.0	30.5
Unoccupied	236	1,613	3.8	35.5
Government Owned	521	12,343	6.5	31.5

basis of how the building is valued should the owner wish to sell it. Income is generated through leases with tenants who occupy the building, which generally include one of these provisions:

- Utility costs, which represent 30 percent of the average building's operating expenses, are passed through to tenants (net lease)

- Utility costs are paid by the owner and calculated into the fixed rent (gross lease)

- Utility costs are locked in over the term of the lease, with the owner paying for increases or benefiting from decreases in energy costs (fixed-base lease)

A typical high-rise building can include dozens, even hundreds, of leases, and many of them may address the subject of utility costs slightly differently.

BENEFITS OF ENERGY EFFICIENCY

If a building owner can reduce its electric operating costs from \$1.06/sq.ft. to \$0.80/sq.ft. through new energy-efficient lamps/ballasts and advanced controls (producing a 50 percent reduction in lighting energy consumption), these benefits can be accrued:

- Net operating income for the building goes up, increasing the building's value (see Figure 11-1). According to the U.S. Environmental Protection Agency (Energy Star Buildings Program), for every $1 invested in energy upgrades such as lighting, asset value increases by $2-3.

- The environment includes more high-quality lighting and other systems designed for occupant needs and is therefore marketable against competitive properties.

- Utility costs are lower, which can be used to attract new tenants.

- Rents can be increased for existing tenants if they are enjoying a demonstrable decrease in pass-through utility costs.

- Direct cost savings benefit in gross or fixed-base leases, increasing the profitability of the lease revenue stream.

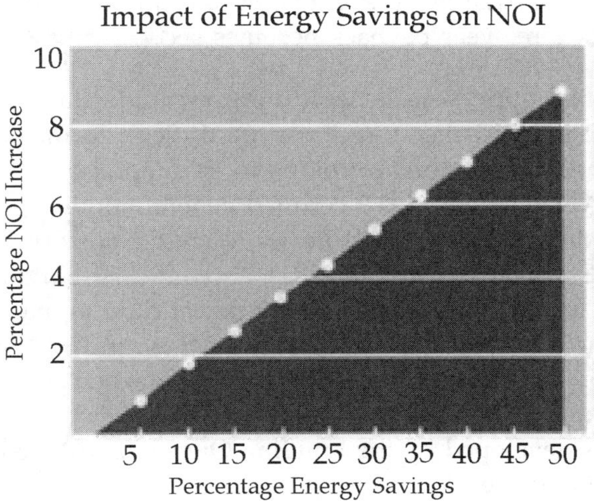

Figure 11-1. Decreasing energy costs improves the net operating income of the property, which increases its value. *Source: U.S. Environmental Protection Agency, Energy Star Buildings Program*

BARRIERS TO ADOPTION OF ENERGY EFFICIENCY

While the above scenario appears to be attractive for both owner and tenant, significant barriers exist to prevent both from taking on the risk of the capital investment:

Owners

- The owner often regards energy efficiency upgrades occurring mid-lease as benefiting only the tenants.

- If a building has dozens or even hundreds of different types of leases, significant administration is required to sort out the cost-benefit and impact on these leases.

- The owner may regard the investment as ideally timed to occur just before the turnover of a lease, which total conversion of its building's lighting systems developing over time based on the tenant turnover rate.

- The vendor of energy-efficient lighting may not understand how the lease is structured before pitching the financial return on the upgrade (for example, if energy costs are split between owner and tenant, a three-year payback becomes six).

- Real estate appraisers generally do not understand energy-efficient design and therefore it can be difficult to include positive cash flow from upgrade projects in the appraisals of real estate value. A survey among 69 certified general appraisers in California conducted by the Institute for Market Transformation found that only 13 percent recognized energy-efficient building features in their appraisals. Nearly half (45 percent) do prepare operating cost schedules, but only 20 percent of these include energy bills. Typically, they use historical income and expense data (59 percent), interviews with owners and sellers (35 percent) or general statistics developed by the Building Owners and Managers Association (43 percent).

Tenants

- The tenant often regards energy efficiency upgrades as benefiting only the owner of the building, even though the remaining period of its lease may be much longer than the typical payback for energy-efficient lighting.

The bottom line in every upgrade opportunity among the commercial lease property market is, "Who pays? Who benefits?"

OVERCOMING THE BARRIERS

The owner generally has a strong incentive to upgrade its lighting systems to benefit both itself and its tenants (which in turn benefits itself). Energy costs in general have been increasing. If the lease is structured so that the tenant bears these increases, the strapped tenant may put pressure on the owner to lower the rent or risk losing the tenant. If the lease is structured so that the owner itself bears cost increases, net operating income erodes with each cost increase, depressing the property's value. As the effects of deregulation, lack of sufficient supply for a stable market, dependence on foreign oil and other factors will bring continued uncertainty to future energy prices, energy will most likely increasingly be a flashpoint in lease negotiation. In today's environment, tenants are more likely to negotiate for leases in which utility costs are fixed.

Administration & Analysis

- Use QuikScope software, developed by the U.S. Environmental Protection Agency, which is designed for commercial property managers. QuikScope, a component of the EPA's Energy Star Buildings Program, helps property owners allocate the costs and benefits of energy performance improvements and determine the financial viability of energy investments.

  For more information:

- Use NOI Builder, proprietary software developed by RealWinWin, Inc., an energy consulting firm specializing in the commercial rental property market. NOI Builder can be used for an investment-grade analysis and modeling to calculate costs, benefits and what-if scenarios.

Figure 11-2. QuikScope, the energy investment performance software from EPA.

Ensuring Higher Property Valuation

- If the owner wants to sell the building, it is not the right time to avoid a capital investment that will increase the building's value.
- If the real estate appraiser does not recognize the value of low energy costs in a valuation, the lending bank usually won't either. In this event, find a lender that will recognize the benefit of energy efficiency in relation to net operating income (and property value) through proper documentation. Documentation includes complete financial analysis (see above) but also complete engineering analysis recognized by a third party such as a reputable engineering firm, Energy Star Buildings Program, Energy Star Benchmarking tool ("Portfolio Manager"), local utility or U.S. Green Building Council's LEED Program.

Example

Building A with $200,000 energy cost savings through energy efficiency measures including electronic-ballasted T8 systems and advanced controls.

	Before Upgrade	*After Upgrade*
Rental Income	$ 20 million	$ 20 million
Energy Costs	$ 3 million	$ 2.8 million
Other Operating Costs	$ 5 million	$ 5 million
Net Operating Income	$ 12 million	$ 12.2 million
Building Value (using 10 percent cap rate)	$120 million	$122 million

- The owner can also split the savings with the tenant in exchange for an increase in rent. For example, the owner can increase the rent by 50-75 percent of the energy cost savings, which are passed along to the tenant. The tenant reduces its electric energy costs by 25-50 percent, while the owner generates an increase in lease revenue. This increase in lease revenue in turn increases the net operating income of the building in a more traditional form accepted by appraisers and lenders.

Example

Building B with $200,000 energy cost savings, passed along to tenants, with 75 percent of the amount added to rent

	Before Upgrade	*After Upgrade*
Rental Income	$ 20 million	$ 20.15 million
Energy Costs	$ 0	$ 0
Other Operating Costs	$ 5 million	$ 5 million
Net Operating Income	$ 15 million	$ 15.15 million
Building Value (using 10 percent cap rate)	$150 million	$151.5 million

With the potential cost savings and added building value, energy efficiency upgrades are often more profitable for investors than riskier speculative investments in new building development.

Cost Savings

- If utility costs are passed along to the tenant, most leases enable owners to recoup these costs before passing through the energy savings.

- If the lease locks in utility costs, the owner keeps the savings.

- Waiting many years for a lease to expire before investing in an upgrade is not the best financial strategy, since there is money on the table today.

Financial Incentives

- New energy legislation currently being reconciled between the Senate and House of Representatives is almost certain to include a tax deduction of up to $2.25/sq.ft. for energy upgrades that exceed the ASHRAE/IES 90.1-1999 energy code by 50 percent.

- More than $1.5 billion in rebates were made available in 2001, more than twice the amount available in 2000, which can be used to reduce the cost of the upgrade.

- Energy service companies (ESCOs) offer guaranteed savings and other performance contracts that start with upgrade financing.

Vendors

- Work with vendors of energy-efficient products and their representatives who understand the commercial real estate market. For example, if energy cost savings are projected to produce a 1.5-year payback but energy savings are split because of the given lease then the payback for the owner is really 3 years. The vendor should be able to produce a complete analysis of the project to help sell senior management and demonstrate that the investment will meet the owner's hurdle return rate.

Chapter 12

Personal Lighting Control: Boosting Productivity, Saving Energy

By Craig DiLouie, Lighting Controls Association

Productivity has traditionally been regarded and measured as work output per man-hour. Today, in non-industrial organizations, productivity is being regarded as a broad range of positive outcomes, with job satisfaction being a leading outcome. Job satisfaction has become increasingly important, particularly for office workers, due to the lengthy period of time required for new employees to reach maximum efficiency, and turnover costs that can erode profitability and competitiveness.

Numerous research studies have shown that workplace design is a major contributing factor to how satisfied and motivated workers are, how well they perform individually, and how they perform as a group. A majority of office workers, however, are not satisfied with the quality of their workplace design, including leading environmental quality factors such as lighting, thermal comfort and acoustics. While people demonstrate highly variable preferences for temperature and light levels, for example, thermal and lighting systems are designed as fixed output systems that will be comfortable for a majority, but not all, occupants.

Since people costs outweigh building costs by a ratio of 13:1, organizations can generate desirable outcomes through investments in productivity, in particular by addressing workplace design. Studies indicate that workers relate comfort to workplace design, and that increasing job satisfaction can correlate to productivity increases.

As a result, organizations today are highly aware of the need for integrating emerging technologies with innovative design to maximize satisfaction and performance among space occupants.

To bridge the gap between a fixed workplace design and highly variable need for lighting and temperature among individuals in a group and for each individual based on changing tasks, time of day and other factors, designers have increasingly adopted personal control solutions. Regarding lighting, this generally entails establishing a dimming system with each occupant in the space being able to interface with the lighting system (through PC, hand-held remote, etc.) to control his or her local light levels.

A number of studies demonstrate that personal dimming can result in higher productivity—specifically in the metrics of vigilance, motivation and satisfaction—and also in energy savings. These advantages are resulting in a significant new trend towards adoption of personal dimming solutions among designers and building owners.

This chapter makes the case for personal dimming.

THE MODERN DEFINITION OF PRODUCTIVITY

Productivity among the complex jobs held in modern offices can be measured in many ways, from forms completed per hour to ideas generated per week, at both the individual and organization level. Today, productivity includes quality of work output, employee attraction and retention, comfort, financial success and job satisfaction. According to the Light Right Consortium this has resulted in an emerging approach to studying worker productivity that focuses on mental building blocks (attention, vigilance, memory, creativity, mental computation, comprehension) and psychological processes (motivation, persistence, effort).

In the Industrial Era, worker productivity was typically measured in the proverbial "widgets per hour," a metric comprised of production output, efficiency and accuracy. In the Information Age and the modern office, worker satisfaction and motivation are now more important metrics due to the complex nature of many office jobs and the high costs of turnover.

According to Harris, Rothberg, LLC, a performance consulting firm, research indicates that the turnover cost for an exempt employee is about 1.2-2 times his or her annual salary. This includes, according to Douglas T. Phillips, author of "The Price Tag of Turnover" (*Personnel Journal*, 1990), inefficiency of the replacement and co-workers working with the replacement; inefficiency of the employee who is leaving and co-workers working with that employee; organizational inefficiency during the time the position is vacant; and processing costs. According to Harris, Rothberg, new employees do not reach maximum efficiency and performance for 13.5 months.

Businesses are sensitive to these costs. In 1999, *Canadian Business* reported that CEOs considered "attracting and retaining high-caliber employees" to be second only to "increasing profitability" as a top corporate priority, ahead of "market expansion" and "mergers and acquisitions."

Job satisfaction may be the key to retaining top employees. A study in the *Journal of Occupational Health Psychology* reported that job satisfaction accounts for 63 percent of variance in organization commitment, which accounts for 80 percent variance in intent to turnover. As stipulated in the study, job satisfaction incorporates satisfaction with the work environment, which brings us to the role workplace design plays in job satisfaction.

In addition to having higher job satisfaction, a more productive office worker demonstrates greater individual and group performance. A more productive office worker performs tasks with greater accuracy, for longer periods of time without tiring, are more creative, can handle stress and unexpected situations better, can interact with other employees more effectively, etc.

WORKPLACE'S RELATIONSHIP TO PRODUCTIVITY

Workplace design has been found to be a major contributing factor to how satisfied and motivated workers are, in addition to how well they perform.

A 1987 study in the *Journal of Applied Psychology* reported that workplace characteristics account for as much as a 31 percent variance in work satisfaction. The Buffalo Organization for Social and Technological Innovation (BOSTI) Associates, an organization that researches the office's effects on productivity and job satisfaction, reported in 2000 that the workplace makes an 8-32 percent (smallest to largest) contribution to job satisfaction (average 24 percent), 3-10 percent contribution to individual performance (average 5 percent), and 6-15 percent contribution to team performance (average 11 percent)—according to a survey of about 13,000 people in 40 business units conducted between 1994 and 2000. And in September 1999, *Sales & Marketing Management* reported the results of a survey of 150 executives, which found that the work environment has become the most important factor in fostering employee satisfaction (see Figure 12-1).

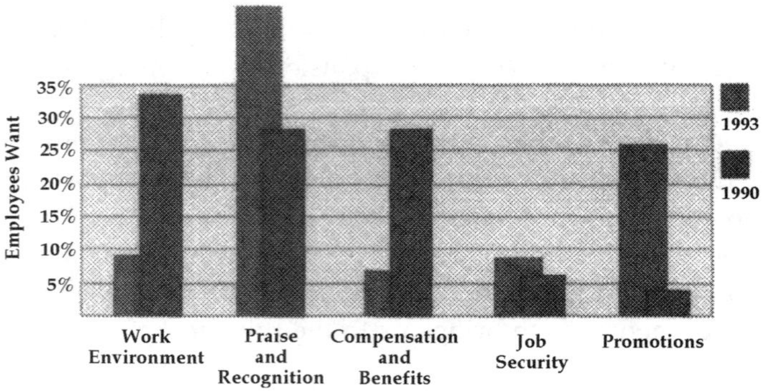

Figure 12-1. A September 1999 Sales & Marketing Management report, based on results of a survey of 150 executives, found that the work environment has become the most important factor in fostering employee satisfaction.

In 1995, The Office of Science and Technology Policy, an arm of the Federal government, stated that better-constructed facilities can result in a "30 percent improvement in productivity and comfort." The Office's *Biennial Report* also stated that better-constructed facilities can result in "50 percent reduction in delivery time; 50 percent reduction in operating, maintenance and energy costs; 50 percent fewer occupant-related injuries and illnesses; 50 percent less waste and pollution; and 50 percent more durability and flexibility."

The 2002 Steelcase Workplace Survey of more than 1,500 corporate executives, facility managers and design professionals from various industries reported that more than three-fourths (79 percent) of respondents believe that "physical comfort has a serious impact on worker satisfaction," while more than one-half (53 percent) believe "their organizations had minimal information regarding the level of satisfaction people have with their physical work environment." This illustrates an alarming disconnect between the organizational goal of productivity and understanding of a key element of that productivity—physical comfort in and satisfaction with the workplace.

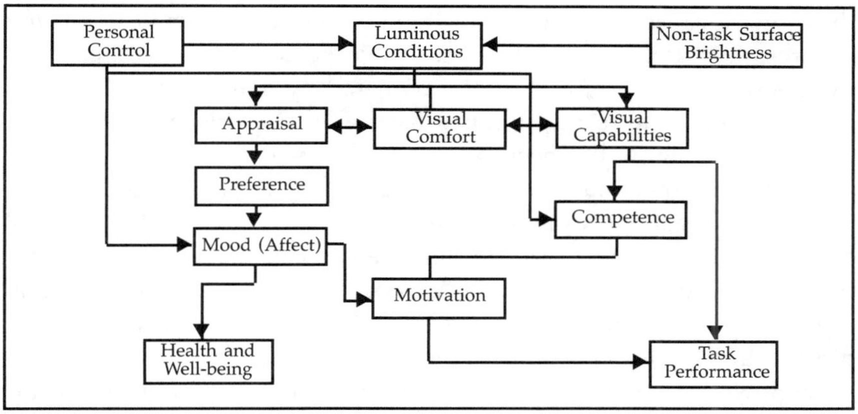

Figure 12-2. How different lighting factors combine to influence worker satisfaction. *Source: Light Right Consortium*

STUDIES FIND WIDESPREAD DISSATISFACTION WITH THE WORKPLACE

While many organizations have failed to connect workplace satisfaction with productivity, the building industry has been unable to satisfy most office workers in regards to thermal comfort, lighting and acoustics. Research indicates that large percentages of workers are not satisfied with their physical workplace.

Besides the actual design of the space (whether it facilitates interactions, communication, ergonomics, privacy, etc.), key elements of the workplace include lighting, thermal comfort and acoustics, which together are components of Indoor Environmental Quality (IEQ). Currently, the American Society of Heating, Refrigeration and Air-Conditioning Engineers (ASHRAE) defines acceptable indoor air quality as "air in which there are no known contaminants at harmful concentrations as determined by cognizant authorities and with which a substantial majority (80 percent or more) of the people exposed do not express dissatisfaction" when temperature is set at 22°C (72°F). Similarly, the lighting industry addressed the issue of glare, which can impair or cause discomfort to vision, and established the Visual Comfort Probability (VCP) rating system, which indicates what percentage of the occupants in the poorest location in the area would not be bothered by direct glare caused by a uniform lighting system of identical lighting fixtures. A VCP of 75, therefore, means 75 percent of the occupants in the poorest location would not be bothered by glare. Generally, office environments require that fixtures have a VCP rating of 70 or more, although this figure has been revised by some in recent years to 80 or more for environments where computers (vertical tasks) are used.

Satisfying the majority is a common-sense approach when thermal and lighting characteristics of the space are fixed. But it also means that the building industry, by design, accepts a one in five dissatisfaction ratio among workers with two key workplace design elements that may directly impact their job satisfaction.

However, even ASHRAE's definition for acceptable indoor air quality and the lighting industry's VCP metric are no guarantee of satisfaction with thermal comfort and lighting for the majority. Studies have found that large percentages of office workers are dissatisfied with thermal comfort, lighting and acoustics in their workplace.

A 1997 American Society of Interior Designers (ASID) study determined that 68 percent of employees complain about the light in their offices. A 1991 Steelcase survey conducted by Louis Harris & Associations discovered that 44 percent of office workers and 64 percent of computer users considered eyestrain (due to glare) to be the leading hazard to their health in the office—ahead of asbestos and even exposure to AIDS. Similar studies document significant dissatisfaction with heating, ventilation and air conditioning (HVAC) and acoustics. A 1983 study by Merck revealed a 43 percent dissatisfaction with HVAC, and 1992 Social Security Administration study found that 56-89 percent of government workers regarded HVAC as a problem. An ASID study found that 70 percent of office workers claim they would be more productive if their offices were less noisy.

The benefits of increased productivity are obvious in terms of individual performance, group performance and job satisfaction. According to a five-year BOSTI Associates study of 6,000 office workers conducted in the 1980s, employment costs exceed building costs by a ratio of 13:1 for owner-occupied buildings and 5:1 for leased space. Therefore, if facility owners, property managers and facility managers can find a way to bridge the gap between building design and worker comfort, they may realize higher productivity and less turnover of employees/tenants. With a 13:1 ratio of people costs to building costs, an investment in building costs to produce even a small increase in productivity can result in a significant impact to the bottom line, helping organizations become more profitable and competitive.

CONNECTING BUILDING SYSTEMS TO INDIVIDUAL NEED: PERSONAL CONTROL

Studies, such as those performed by Dr. David Wyon of the National Institute of Occupational Health in Copenhagen, Denmark, have demonstrated that people respond very differently to their environment. Wyon showed that workers who are satisfied with their environment are up to 15 percent more productive compared to people who are not.

Numerous studies have shown that workers are more satisfied with their working environment when they have control over thermal comfort and lighting. Personal control bridges the cap between a building design that attempts to satisfy the majority and people who have very different needs based on a range of factors.

Thermal Comfort

Wyon estimates that group performance can realize an improvement of 2.7-8.6 percent by providing individual control over the environment rather than trying to reach a temperature acceptable for all or most workers. Carol Lomonaco and Dennis Miller of Johnson Controls, in an important white paper, "Environmental Satisfaction, Personal Control and the Positive Correlation to Increased Productivity," write, "When office workers are satisfied with their environmental conditions, when they can work in greater comfort and control, they will be more productive. Additionally, the cost of employment per worker will drop, and the cost of facilities operation will decrease... a growing body of research supports these conclusions."

Lomonaco and Miller cite several studies supporting the hypothesis that individual control of thermal comfort leads to greater productivity. In a study conducted at the University of California in Berekely and a similar study conducted in 12 air-conditioned offices in Townsville, Australia, occupants exhibited a wide range of preferences for thermal conditions. The Office of the Environment Study (United Kingdom) examined several variables and their effect on productivity, including number of people in a given room and job type, and level of personal control. The study found that productivity decreased as the number of people in the room increased. The study also found that productivity increased as the level of personal control increased, independent of the number of people in the room.

The Johnson Controls authors provide a clear, comprehensive and persuasive argument for individual control of thermal conditions in offices, backed by a considerable amount of research and references. The rest of this chapter will explore the correlation between personal control over light level and productivity.

Table 12-1. Analysis of Environmental Satisfaction-Productivity Studies. *Source: "Environmental Satisfaction, Personal Control and the Positive Correlation to Increased Productivity," Carol Lomonaco and Dennis Miller, Johnson Controls, Inc.*

Analysis of Environmental Satisfaction-Productivity Studies

Study	*Environmental Condition*	*Result*
Greening the Buildings & Bottom Line 1994	Lighting Noise & Daylighting New Building	6 percent prod. gain, reduced defects, $25k increase in product quality, 13 percent prod. gain, 25 percent less absenteeism 15 percent prod. gain + 15 percent less absenteeism 15 percent less absenteeism
West Bend Mutual 1992	Individual Control	2.8 percent prod. gain/could be up to 6 percent 12.8 percent prod. drop when disconnected
Mau-Lin Chiu/Carnegie Mellon 1991 According to Chiu, six factors influence office productivity: (1) Spatial Quality (2) Thermal Quality (3) Visual Quality (4) Acoustic Quality (5) Air Quality (6) Long-Term Building Integrity	Lighting Noise Temp & Air Quality	Cites 4 Studies Cites 5 Studies Cites 5 Studies

Table 12-1. (Continued

Analysis of Environmental Satisfaction-Productivity Studies

Study	*Environmental Condition*	*Result*
Economic Benefits of a Healthy Indoor Environment (Wyon) 1994	Thermal Air Quality Individual Control	5-15 percent incr. efficiency in concentration 34 percent improvement in Sick Building Syndrome
Predicting the Effects of Individual Control on Productivity (Wyon) 1995	Individual Control	3-25 percent efficiency gains 3-15 percent for concentration and 7-25 percent for routine office tasks
Indoor Air '96 Conference (Wyon)1996	Individual Control	2-10 percent increase in group efficiency
BOSTI 1984	Noise Temperature/Air Quality Lighting Comfort	These each have dollar figures for 3 job types representing improvements to absenteeism and turnover.
Center Core 1993	Air Quality	55 percent improvement in absenteeism 40 percent self-reported prod. increase Enhanced perception of their office
UK Office of Environment 1990	Air Quality Space (People per room) Individual Control	These are all self-reported results. People feel they are more productive when air quality is better, they are less concentrated, and have environmentalcontrol.
Worker Productivity: Hidden HVAC Cost 1990	Air Quality	One lost working day due to a sick building = 60 percent of annual energy costs.
Lighting & Human Performance (refers to other studies)	Lighting	7.6 percent prod. increase 13.2 percent prod. increase 30 percent proofreading decrease when light levels cut in half.

Lighting

People vary significantly in their preferences of lighting. Age is a significant factor in how much light an individual needs to perform a given task accurately and efficiently. It has also long been known that various tasks demand different light levels depending on contrast, size and time allowed for the task. Depending on location, workers may be forced to suffer from the effects of glare, which causes eyestrain, which in turn is considered a leading health hazard by many office workers. Daylighting can be beneficial but the lighting system must be able to respond to changing light levels to save energy and eliminate glare. In addition, workers today are expected to perform a greater variety of tasks in the same space, use computers (vertical in addition to horizontal tasks), and handle greater workloads that previously had been the responsibility of a larger workforce. To accommodate these working conditions, they need optimum lighting conditions perfectly tuned to their needs—essentially, the ability to tune their lighting according to changing tasks, mood and ambient conditions (such as time of day and amount of daylight).

Personal lighting control satisfies these needs. It has been demonstrated in numerous studies to increase job satisfaction, motivation, vigilance and performance—by bridging the gap between a fixed building design and a highly variable individual need. Advancements in lighting technology now provide cost-effective personal control capabilities to buildings that can improve productivity as well as energy savings.

PERSONAL DIMMING CONTROL: RESEARCH STUDY #1

According to the California Energy Commission, automatic lighting controls generate typical energy savings of 35-45 percent in commercial and institutional buildings. Personal dimming control in private offices can accelerate energy savings while increasing occupant satisfaction and enhancing the value of the space.

The first major research in this area, conducted by the Lighting Research Center, demonstrated manual dimming energy savings of 6 percent in its eight-week study of 58 private offices at the National Center for Atmospheric Research (NCAR), a three-building, 250,000 sq.ft. complex in Boulder, CO.

Each office was lighted with two 2x4 recessed troffers housing three 32W T8 lamps driven by dimmable electronic ballasts. The lighting controls included a wall-mounted manual unit for on-off and dimming; a portable manual dimmer on the desktop; and a PIR occupancy sensor mounted in a corner for automatic switching.

The Lighting Research Center reported energy savings of 61 percent, with 43 percent from occupancy sensors, 6 percent from manual dimming, and the rest from other methods. Three out of four of the occupants used the manual dimmers at least once and used the desktop dimmer over the wall-mounted unit by a ratio of six to one. The occupants also used their manual controls to switch the lights and work under daylight entering the room through window blinds.

The biggest reason they dimmed their lights? Computers, they said. "Compensation for daylight," "read printed text," and "create an atmosphere for work" were other important reasons to 10-20 percent of the survey participants.

Whatever their specific reasons, the Lighting Research Center concluded, "Employees... prefer manual lighting control to automatic controls because the manual controls allow them to tailor the lighting to their needs."

"Employees like to have control over their work environment," says A.J. Glaser, president of HUNT Dimming and the Lighting Controls Association. HUNT Dimming provided equipment for the NCAR research project. "Using manual dimming devices gives occupants the chance to tune light levels according to their preferences and needs, which increases their satisfaction while saving energy."

PERSONAL DIMMING CONTROL: RESEARCH STUDY #2

In 2002, National Research Council (NRC) Canada published a research study, "Preferred Surface Illuminances [light levels] and the Benefits of Individual Lighting Control: A Pilot Study," authored by Guy R. Newsham, C. Arsenault and Jennifer A. Veitch. The researchers established two different lighting conditions in two workstations in a mock-up open-plan office. One was fitted with conventional ceiling-recessed parabolic lighting fixtures that were dimmable (A). The adjacent workstation was fitted with a dimmable "partion-washer" system designed to preferentially light vertical surfaces in the worker's view, supplemented with light from overhead fixtures (B).

Twenty-two participants, all of them lighting experts, performed a task in one of the two spaces and then completed a questionnaire about their satisfaction with the lighting. They then dimmed the lighting to their own preference using an interface on their PCs and repeated the task and the questionnaire. After that, they switched workstations and repeated the entire process.

The researchers observed that the participants preferred a wide variety of light levels. While preferences clustered at around 19-28 foot-candles (fc) (200-300 lux) on the desktop in Workstation A, preferred light levels overall ranged from 5-84 fc (50-900 lx). While preferences clustered at around 28-37 fc (300-400 lx) on the desktop in Workstation B, preferred light levels overall ranged from 5-74 fc (50-800 lx).

"While we would expect a wide range in the preferred luminous conditions produced by individuals, we would predict broad agreement that control, not matter how it is used, is beneficial," noted the authors. "On average participants agreed that their own lighting choice improved their ability to do the job well compared to the lighting they started with... These positive effects associated with individual control and receiving preferred lighting conditions are expected, and agree with other recent research work on individual lighting control."

PERSONAL DIMMING CONTROL: RESEARCH STUDY #3

The most significant research about the effects of personal dimming control was conducted by the Light Right Consortium. The Consortium's landmark study, formed to address the benefits of quality lighting, indicates that personal control of lighting can result in a significant improvement in occupant satisfaction and performance.

The Light Right Consortium's goal is to transform the lighting market by using research to investigate the link between lighting quality and the performance, satisfaction and productivity of workers. The Consortium, formed in 1998, is managed by the Pacific Northwest National Laboratory and operated by Battelle for the U.S. Department of Energy. Board members include the Alliance to Save Energy, the Illuminating Engineering Society of North America, the International Association of Lighting Designers, the International Facility Managers Association, Johnson Controls, the National Electrical Manufacturers

Association, the New York State Energy Research and Development Authority, Steelcase, the U.S. Department of Energy and the U.S. Environmental Protection Agency. Project sponsors who contributed equipment included Armstrong, Birchwood Lighting, Cooper Lighting, Day-Brite Lighting, Engineered Lighting Product, General Electric, Ledalite, Lightolier, Lutron, OSRAM SYLVANIA, Peerless Lighting and Philips Lighting. The Lighting Research Center and the National Research Council of Canada were contracted to perform the research.

Figure 12-3. The most significant research about the effects of personal dimming control was conducted by the Light Right Consortium at this Albany, NY, mock-up office. The Consortium's landmark study, formed to address the benefits of quality lighting, indicates that personal control of lighting can result in a significant improvement in occupant satisfaction and performance.

"Central to the success of the Consortium is establishment of a link, based on sound research results, between quality lighting and economic benefits," says Carol C. Jones, LC, Program Manager. "Market transformation goals include 1) influencing customer decisions so that they are designing, purchasing and installing higher-quality and more energy-efficient technologies, 2) going beyond the technology issues to delve into the dynamic of customer and market behaviors, and 3) creating enduring market changes."

First, the Consortium conducted market research to provide proof of concept. A survey was conducted among professionals who specify, install and own/use lighting systems. It was not surprising that 87 percent of respondents reported flexibility in lighting budgets if a return on investment could be demonstrated. But 75 percent said if factual evidence indicating a positive effect by lighting on worker productivity was available, it would influence which lighting systems they would buy. These results validated the need for Phase II, which was to provide this factual evidence validated by scientific method, and to study which lighting approaches were the most effective at influencing workers.

With Phase II's implementation, a research program was formulated to address the question, "Can different forms of realistic office lighting affect the performance of office work or the well-being of employees?" The primary variables included room surface brightness and personal control. The study indicated that personal dimming control resulted in occupants performing better on certain productivity metrics.

An office in Albany, NY, was set up as a typical space for nine workers. The open office plan featured perimeter windows and access to a view, although translucent window shades were used to alleviate the impact of daylight at workstations. The space was planned and furnished to allow the researchers to change the lighting between five different lighting systems without the knowledge of the subjects. The workers were temporaries hired to work under the different lighting conditions for a typical eight-hour day. A range of output measures were collected that ranged from the subjective (occupant opinion) to objective (quantitative performance), resulting in a large data set. The study was conducted in the field, but with simulated tasks and a degree of experimental control. This approach was chosen to maximize realism and the validity of the research.

The four lighting scenarios included:

- "**Best Practice**": Linear system of direct/indirect fixtures together with wall-washing to brighten the walls.

- "**Switching Control**": The same as best practice but with a moveable desk lamp having three manually switched light outputs and providing some individual control.

- "**Dimming Control**": Direct/Indirect fixtures suspended over the center of each cube, together with wall-washing system. The direct component of each could be dimmed using the interface on the occupant's computer.
- "**Parabolic Base Case**": Regular array of three-lamp parabolic-louvered fixtures.
- "**Lensed Troffer Base Case**": Regular array of recessed lensed troffer fixtures.

The temporaries worked for a complete day on set tasks to simulate elements of office work, and on questionnaires linked to the productivity metrics being studied.

When asked whether they agreed with the following statements at the end of the day, the workers responded:

"Overall, the lighting is comfortable."

Direct/Indirect with Dimming Control	91 percent
Parabolic Base Case	71 percent

"The lighting is uncomfortably bright for the tasks that I perform."

Direct/Indirect with Dimming Control	11 percent
Parabolic Base Case	33 percent

"The lighting causes deep shadows."

Direct/Indirect with Dimming Control	12 percent
Parabolic Base Case	7 percent

"The lighting fixtures are too bright."

Direct/Indirect with Dimming Control	19 percent
Parabolic Base Case	38 percent

"Reflections from the light fixtures hinder my work."

Direct/Indirect with Dimming Control	29 percent
Parabolic Base Case	21 percent

Personal dimming control with linear suspended direct/indirect fixtures yielded a 30-point spread in response to whether the workers believed the lighting was comfortable, and produced the lowest incidence of workers perceiving their lighting to be uncomfortably bright for the tasks they performed. Participants were also asked:

"How does the lighting compare to similar workplaces in other buildings?"

	Worse	*Same*	*Better*
Direct/Indirect with Dimming Control	7 percent	43 percent	50 percent
Parabolic Base Case	8 percent	69 percent	24 percent

In the objective segment of the research, the Light Right Consortium discovered that the presence of control had a measurable impact on motivation, which in turn was represented in the study in measures of persistence and vigilance.

The Consortium concluded:

"People with dimming control reported higher ratings of lighting quality, overall environmental satisfaction, and self-rated productivity... people

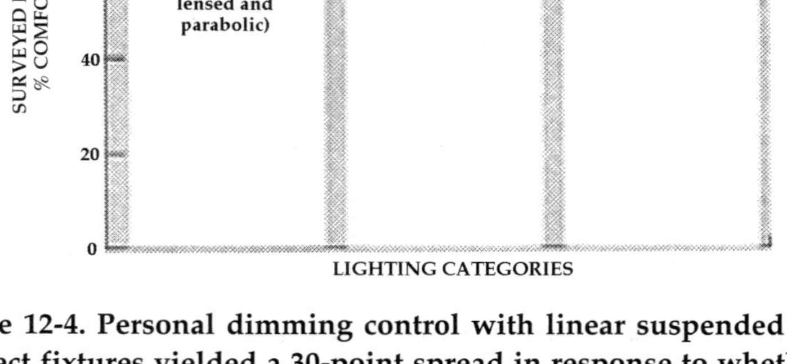

Figure 12-4. Personal dimming control with linear suspended direct/indirect fixtures yielded a 30-point spread in response to whether the workers believed the lighting was comfortable, compared to the baseline case of parabolic fixtures. *Source: Light Right Consortium*

with dimming control showed more sustained motivation, and improved performance on a measure of attention... In addition, on average, people with dimming control chose lower illuminances [light levels] than current recommended practice. This implies that individual overhead dimming control has potential for energy savings."

Persistence at a difficult or impossible task is an indicator of motivation at the task; people who are not motivated to do the task will not continue at it when it becomes very difficult. Vigilance is a state of watchfulness or careful attention, and is related to accuracy. The study subjects were more able to sustain their persistence and vigilance over the day in the personal dimming scenario compared to the baseline and best practice conditions. The probable reasons for this included:

- The ability to fine-tune the lighting conditions to meet the needs of individuals, both with respect to horizontal light levels and the brightness on the surrounding partitions.

- The ability to satisfy the preferences of individuals—the function of satisfaction in the workplace.

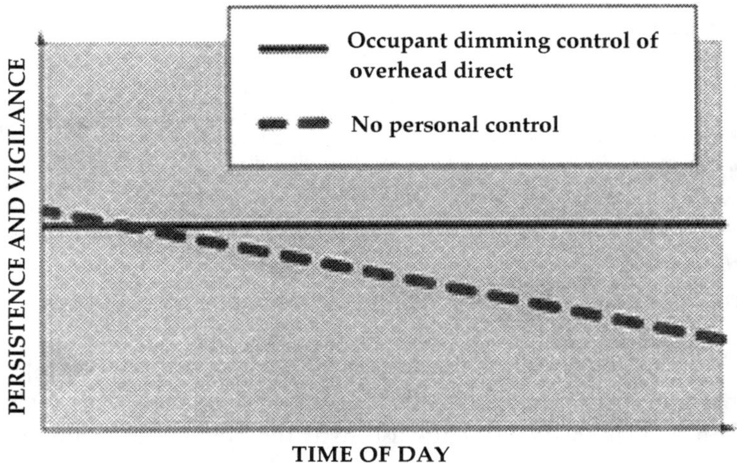

Figure 12-5. In the objective segment of the research, the Light Right Consortium discovered that the presence of control had a measurable impact on motivation, which in turn was represented in the study in measures of persistence and vigilance.

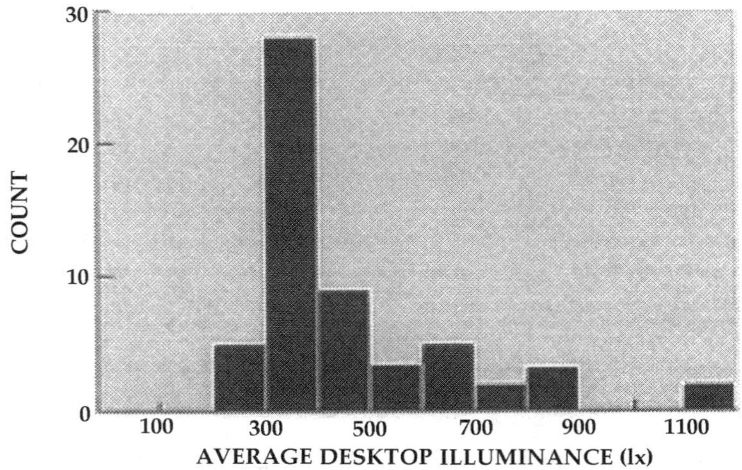

Figure 12-6. The Light Right Consortium study confirmed the findings of previous studies, which indicated that people have a wide variety of light level preferences.

- The psychological impact of control on motivation.

  The study concluded:

  "Dimming control participants showed steeper performance improvements over increasing contrast in the timed vision task and avoided motivation declines over the day. They also improved in vigilance performance over the day, whereas the Best Practice participants did not. There was additional evidence in interaction effects with Print size and time that typing performance also showed beneficial effects of having dimming control."

"Perhaps the simplest and most profound message with respect to personal control is that we are learning that personal control significantly improves our ability to optimize the satisfaction and performance of office workers," says Jones. "We know from prior work conducted at National Research Canada that it there is a great variety of preferred light levels. This tells us that we have a tremendous opportunity, and a tremendous challenge, if we choose to raise the bar with respect to meeting the needs of the office worker population."

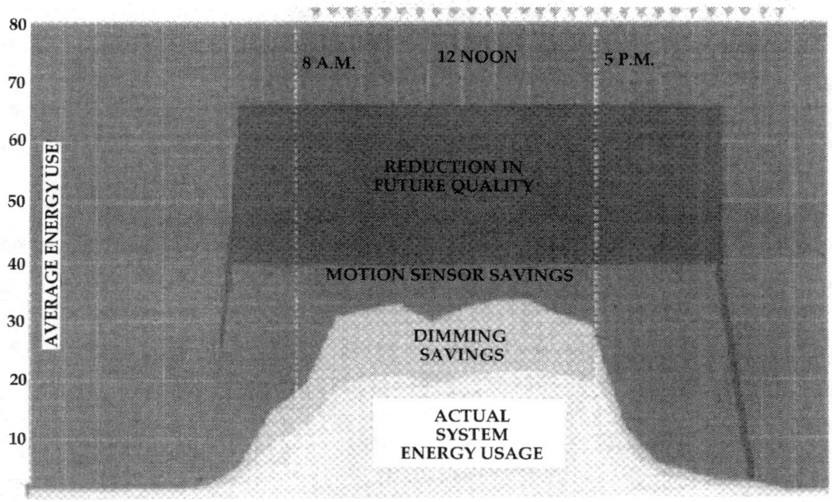

Figure 12-7 Personal control can increase energy savings in addition to other control strategies and fixed load strategies. *Source: Light Right Consortium*

PERSONAL DIMMING: A NEW TREND

Facility owners are becoming aware of the benefits of personal lighting control and are willing to pay a premium as an investment in worker satisfaction and performance as well as a new source of energy savings.

Below is a qualitative look at how some executives are regarding the benefits of personal dimming control, collected by Lutron and printed in its literature:

> *"Our employees are better off with improved lighting. And they like personal control of their lighting environment. So there's no doubt in my mind that they are more satisfied and productive. And I'd say we have cut our lighting costs by 50 percent, saving us hundreds of dollars per month."*
>
> *—John Tomczak, President, Pro-Tech Industries*

> *"Our fluorescent dimming system has met our objective of providing personal control and employee comfort. People are using the system;*

some change light levels as they go from task to task, others leave lights at a low level throughout the day. But you'll observe that every person has a different light level."

—Ray Bromfield, Project Manager, Facilities, America Online Incorporated

And below, as reported in the February 2002 issue of Today's Facility Manager, a leading lighting designer describes an experience with personal dimming:

"We have done several projects for a local university. As a standard, we provide lighting intensity controls in each open and enclosed office. Workers can increase or decrease the light level in their personal space. The university's facilities management department has gotten a great deal of positive feedback on this arrangement. People feel more productive because they have more personal control over their own environment."

—Alfred Borden, IALD, Principal, The Lighting Practice

In 2003, The Watt Stopper, a controls manufacturer, commissioned a study conducted by Ducker Research, which consisted of telephone interviews of 158 facility managers, electrical engineers and architects. The study found that lighting automation is becoming the norm rather than the exception in new construction. It also determined that providing personal dimming control to occupants is gaining acceptance.

The study asked respondents to rate factors driving the use of automated lighting controls. "Providing occupant control capability" ranked fourth in the top five, after "increasing energy savings," "complying with owner requests," and "compliance with state and national energy codes." It ranked above "obtaining utility rebates and incentives."

The study then identified five trends influencing the controls field and asked respondents to rate each trend on a scale of 1-5, from extremely important (1) to not important (5). "Increased need for enhanced occupant control of lighting" ranked third, after "standard protocols for lighting automation systems" and "integration of lighting automation system with the building management system." It ranked above "increased demand for flexible use of space" and "increased use of architectural daylighting design practices."

After being identified as a major trend, occupant control was at-

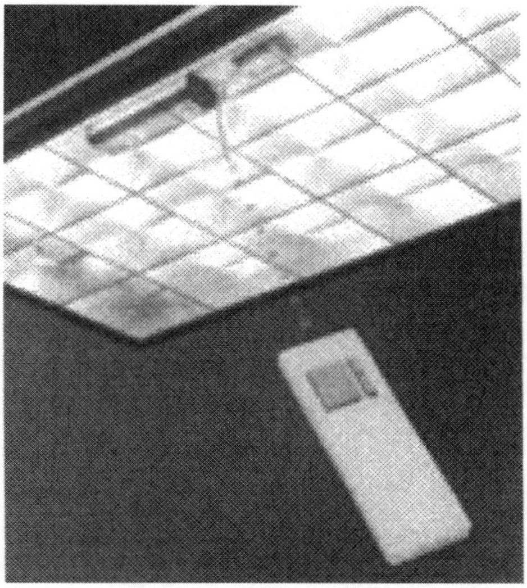

Figure 12-8. Personal dimming control devices. This system uses an RF wireless hand-held remote to achieve dimming. *Source: Lutron Electronics*

tached a cost. A choice was provided: Given the installed cost for a traditional parabolic system is $2.00 per sq.ft., which of the following three options would they elect to use to improve lighting quality?

#1	Use a direct/indirect fixture for $2.50/sq.ft. installed	40.3 percent
#2	Integrate occupancy sensors for $3.00/sq.ft. installed	31.3 percent
#3	Integrate occupancy sensors and provide personal dimming control for $3.50/sq.ft. installed	25.4 percent

Option #1 was desirable to respondents primarily because it represented a lower initial cost. Option #2, however, was desirable primarily because it is "cost effective, a good value." Option #3 was desirable primarily because it increased occupant comfort. The implication of the positive response to personal dimming control is that a significant segment of the market would pay a premium of $0.50 per sq.ft. for it.

Chapter 13

Good Controls Design Key to Saving Energy with Daylighting

By Craig DiLouie, Lighting Controls Association

Daylighting has become a more important feature of mainstream construction due to the sustainable design movement. Daylighting is the use of daylight as a primary source of illumination in a space.

"Spaces that are daylit provide an improved sense of well being," says Chris Meek, Research Associate for the Daylighting Lab and a Lecturer in Architecture at the University of Washington. "Increased access to daylight versus no daylight in classrooms has been correlated to large increases in student test scores. Similarly, extremely large increases in retail sales have been attributed to the illumination of grocery stores with daylight via skylights."

These and other studies have been conducted by the Heschong Mahone Group and illustrate significant potential benefits of daylighting in commercial spaces.

"Many studies over the last 50 years have shown that workers prefer to have daylight and views in their work space," he adds. "When looking at the bottom line, owners need to recognize that 80-90 percent of their costs are often in staff salaries and benefits. Anything that enhances staff performance pays back at an enormous rate, and if you are careful, the project can save a great deal of energy."

These productivity and energy savings benefits have been recognized by the U.S. Green Building Council's LEED Rating System, Lightfair International (which launched a specialized education program), and the 2005 version of California's Title 24 energy code.

Daylighting and LEED

Lighting is related to achieving at least 8 points and as many as 22 points in these sections: Sustainable Sites, Energy & Atmosphere, Indoor Environmental Quality, and potentially Innovation & Design Process. Daylighting, which intersects with LEED requirements in Indoor

Environmental Quality, Credit 8.1: Daylight and Views, can earn 1 point. This credit requires that 75 percent of all critical visual task-occupied space must achieve a daylight factor of 2 percent, and occupants in 90 percent of regularly occupied spaces must have a direct line of sight to vision glazing.

Daylighting and California's Title 24

"The 2005 Title 24 is the first instance of skylights being required by a building code for energy savings," says Lisa Heschong, Principal of the Heschong Mahone Group.

Title 24 now identifies skylights combined with daylighting controls as the baseline efficiency standard for big box-type spaces.

A prescriptive provision requires skylights in these big box buildings, specifically skylights with controls to shut off the lights when daylight is available. The provision applies to buildings >25,000 sq.ft. with >15 ft. ceilings, and to spaces directly under a roof and with general lighting power density of >0.5W/sq.ft. For these spaces, at least one-half of the floor area must be daylit using skylights. The skylights must have a glazing material or diffuser that effectively diffuses the skylight.

In addition, for daylit areas larger than 250 sq.ft., at least one control is required to either control 50 percent of the power, control fixtures in vertically daylit areas separately from horizontally daylit areas, or maintain uniform levels by means of dimming or alternating lamp/fixture switching. For daylit areas over 2,500 sq.ft., general lighting has to be controlled separately with either automatic multi-level daylighting control or multi-level astronomical time switch, both having to meet requirements of Section 119 (Section 131, 143).

Seventy-five percent of all commercial space in the United States is one-story and directly under a roof, representing significant potential for skylights in office, school, gym, retail, warehouse and similar buildings.

For a daylighting strategy to be successful, the designer must effectively design the electric lighting system.

Electric Lighting Design

"The key is to have an overall vision of how the lighting system and lighting controls integrate with the daylighting scheme," says Doug Paton, Daylighting Product Line Manager for The Watt Stopper.

"Lighting is a tremendously important architectural element that has far-reaching impacts on energy consumption, operating costs, aesthetics and ambiance, user satisfaction, worker productivity, and the environment," says Stuart Berjansky, Senior Product Manager, Dimming for Advance Transformer Company. "Lighting is often relegated to low-priority status within the design/build process, when in reality it should be considered upfront and incorporated into the entire building design for maximum effect and benefit."

In short, electric lighting and daylighting systems should be designed so that they are integrated and complementary. For example, when warm color temperature fluorescent sources (<3500K) are used in conjunction with very cool daylight (>5000K), the lights may appear yellow. To mitigate this effect, many designers specify lamps with a neutral-white color temperature of 3500-4100+K.

Controls are a major area of integration. Daylighting entails the use of daylight as a primary source of illumination. Since daylight is generally available during hours when most commercial buildings are occupied, daylighting is often feasible. However, if the lights are operating at full output when there is ample daylight available, then no energy is being saved and the owner is wasting money. If the building is heated or cooled, daylighting may even result in a net increase in energy consumption if daylighting controls are not present.

"Demand for daylighting controls continues to increase as more buildings are designed for sustainability," says Paton. "Lighting controls make a daylit space an energy-saving opportunity."

The strategy is called daylight harvesting. "To some lighting enthusiasts, daylight harvesting may mean use of some active and dynamic method of increasing the quantity of daylight entering a building, but these applications are still rare," says Pekka Hakkarainen, Ph.D., Director, Technology & Business Development for Lutron Electronics Co., Inc. "There are emerging technologies in skylights and controllable louver systems that provide such active dynamic control. More commonly, daylight harvesting means simply making use of daylight and reducing electric light intensity in the building. These applications are seen in many public buildings, educational facilities and higher-end office buildings."

According to Heschong Mahone, energy savings from daylighting controls can range from about $0.5/sq.ft. to $0.75/sq.ft., depending on the building type, location, operation and local cost of energy. Primary factors include the amount of daylight available and the occupancy pattern, plus the control strategy. In addition, since ample daylight is often available during utility peak demand hours (usually 3 to 6 pm), daylight harvesting can reduce demand charges, particularly valuable if a "ratchet clause" is in effect.

Using Controls to Integrate Lighting and Daylighting

There are basically four options available:

- Manual dimming. Occupants can be given the capability of dimming the lights in an area. However, this will probably not result in maximum energy savings.
- Automatic shut-off. This can be accomplished using one of two methods. The simplest method is to use a timeclock. On a regular schedule, the entire fixture can be shut off or individual lamps can be shut off to achieve dual light levels, typically 100 percent and 50 percent. The other method is use a light sensor combined with a relay and switch. The light sensor measures ambient daylight and if enough light is measured, the fixture or individual lamps within the fixture are switched off. Staging the switching in a fixture by enabling shut-off of, say, two lamps, then the other two lamps, is often called stepped switching.
- Automatic stepped dimming. Similar to automatic shut-off, stepped dimming can be based on a time-of-day schedule or on sensed quantity of daylight. However, with dimming, light output is gradually reduced, which is less jarring than lights switching on and off. Stepped dimming is often called bi-level dimming because the strategy often involves two levels of light output, usually 100 percent and 50 percent. However, if more flexibility is required, stepped dimming can involve three levels of light output.
- Automatic continuous dimming. Based on a schedule or sensed quantity of daylight, fixture light output can be gradually dimmed over the full range, from 100 percent to 1/5/10 percent (fluorescent) or 100 percent to 50 percent (HID).

Choosing the right strategy depends on the application requirements.

Figure 13-1. When the 1920s era building at Montgomery Park in Baltimore was renovated to feature state-of-the-art lighting, Sam Himmelrich, Jr. of Himmelrich Associates, the property developer, opted for a daylight harvesting system using photocells. The sensor essentially sends a signal to the local lighting system, telling it how much dimming to engage based on available daylight levels. "Continuous dimming with daylight trackers was an effective and affordable approach at Montgomery Park," says Himmelrich. "The system works, it's straightforward and simple for end-users to operate, and it minimizes energy use and maintenance requirements." Advance Transformer supplied more than 2,000 Mark VII dimming ballasts for the project, in addition to some 1,300 Centrium non-dimmable electronic ballasts for areas where photocells were not installed. The building's first tenant? The Maryland Department of the Environment.

Manual Dimming

Manual dimming is a simple strategy but does not provide programmable control and generally does not provide maximum energy savings because the level of control is dependent on the vigilance of the user. In a manual dimming strategy, the user should have control over their immediate work area via an easily accessible control device. Individuals with responsibility for larger spaces, such building managers, should have authority to control larger control zones. Even if an automatic control strategy is chosen, local user manual override may be desirable.

Automatic Shut-off

Automatic Shut-off can provide a low-cost option and is suitable for applications where there is ample daylight that is highly predictable. However, if the entire fixture is shut off, occupants may complain about interruptions (lights suddenly activating and de-activating for no understood reason) and lights being off (again for no apparent reason). In these situations, it is often good practice to educate users about how their control system works, and that their building is using this strategy to save energy.

"Good daylighting control will not annoy occupants," says Paton. "In fact, if designed correctly, daylighting control has the ability to delight occupants."

Stepped Dimming

Stepped dimming is popular for HID lighting systems as a lower-cost option for spaces where full light output is needed quickly after switching on (as HID Lamps require a warm-up and restrike time). Stepped dimming is also suitable for fluorescent systems in spaces where daylight levels are variable, where ample daylight is not predictable. In addition, stepped dimming is often considered desirable for spaces with major motion activity such as walking and shelf stocking.

Continuous Dimming

Because continuous dimming follows the daylight pattern very closely, it not only is often more acceptable to occupants, but can produce higher energy savings, particularly in areas with highly variable cloud cover. Continuous dimming also responds to changes in light output due to dirt depreciation on fixtures and lamps, and lamp lumen depreciation due to lamp aging. Lighting systems are generally overdesigned to compensate for these light loss factors, with an initial light output that is typically 15-25 percent higher than at end of life. By maintaining a constant light level, dimming can compensate for this overdesign and increase energy savings. According to Heschong Mahone, lumen maintenance dimming can result in an additional 5-10 percent energy savings over the life of the lamps.

Continuous dimming also provides the highest degree of flexibility, particularly for spaces where daylight levels are variable during the day. In addition, continuous dimming can provide greater uniformity of light levels in daylit spaces where some areas receive lower amounts of daylight than others. Continuous dimming is often considered desirable for spaces with minor motion activity, such as reading, writing and conferencing—such as offices and classrooms.

Dimming is often considered to be better design practice in terms of occupant perception. When lamps are switched, the sudden change of light output is noticeable to occupants, and the occupants are suddenly being told they have less light, which can be irritating. When lamps are dimmed, light level decreases but the human eye may perceive a higher light level than is actually recorded by a light meter. For example, 1 percent measured light is actually perceived as 10 percent, 5 percent as 22 percent, and 10 percent as 32 percent (following the square law). In addition, research conducted by the Lighting Research Center suggests that people do not notice changes in light levels from dimming as much as they do from switching.

When designing a continuous dimming system, an important consideration will be the creation of control zones. All lamps in a given control zone are dimmed and regulated by a controller and a photosensor. For gaseous discharge lighting systems, each lamp is dimmed using a dimming ballast.

Continuous dimming is achievable using either analog or digital ballasts.

Analog ballasts are currently the most popular type and may be 0-10VDC, two-wire phase-control, three-wire phase-control or wireless infrared.

Digital ballasts are a more recent introduction and offer new opportunities in dimming and lighting control. Benefits include greater granularity of lighting control, reconfiguration without rewiring, possibility of providing an estimate of energy consumption, and smaller control zones than was previously practical. Most digital ballasts are compatible with the DALI protocol.

"A major technological trend that is positive for the industry is the continued drive towards cross-compatibility among various controls manufacturers," says A.J. Glaser, President of the Lighting Controls Association and HUNT Dimming.

Controls Application Tips

When planning a controls system to integrate with a daylighting strategy, consider the following tips:

- Integrate lighting and lighting controls design into the initial planning and design process.
- Design control circuits parallel to the daylight contours to create control "zones" that match daylight availability/coverage patterns.

- Allow users to manually override automatic controls.
- Consider integrating automatic lighting control with automatic window shades, blinds or other devices that can reduce direct glare and heat gain.
- Adapting existing buildings for dimming is often not difficult, but adapting them to daylighting can be. Single-story buildings with simple roof structures are often easiest to upgrade for daylighting, particularly spaces with high ceilings.
- Specify commissioning services as a separate item, to be bid separately.
- Light sensors should be located carefully to synchronize the availability of daylight with coverage from the electric light fixtures. "The location of the photocell should be indicated on the bid documents," says Paton. "Unfortunately, the mounting requirements are manufacturer-specific. It is crucial to understand that a location that is selected based on the recommendations of one manufacturer may not work on another manufacturer's product."
- Write a sequence of operation for the lighting controls. "This is a great tool for clearly communicating the intent of the controls system design," says Paton.
- Provide specific guidelines and expectations for checkout and verification of the lighting controls.
- Demand performance specifications from the controls manufacturers. "Carefully read and follow the photocell [light sensor] placement guidelines in your designs."

"New daylighting controls that measure light in the same way that people perceive it will significantly improve daylighting control," says Paton.

"The commissioning process has been simplified in the last several years, and manufacturers have trained technicians who know how to perform this job in a high-quality fashion," says Hakkarainen.

"If daylighting is done correctly, from specification of complementary equipment to proper installation and commissioning, it works," says Glaser.

Chapter 14

2005 NEC Changes Impact Lighting Control Panels, Metal Halide Lighting

By Craig DiLouie, Lighting Controls Association

Nearly all 50 states rely on the National Electrical Code (NEC), published as a standard by the National Fire Protection Association (NFPA), as the code they use to regulate electrical installation in new construction and renovation projects. The NFPA recently published the 2005 version of the NEC, which is enforceable (at the state and/or municipal level) in all states and municipalities that adopt it. Several provisions in the new Code affect lighting, including lighting control panels, metal halide fixtures, and disconnecting fluorescent fixtures prior to servicing. The 2005 NEC went into effect January 1, 2005, but adoption can vary from state to state. Based on the rate of adoption of the 2002 NEC, several states may adopt the Code right away, with about half the states adopting it in 2005. At the time of writing, according to the National Electrical Manufacturers Association (NEMA), several states, including Texas, North Carolina and New Hampshire, have adopted the 2005 NEC, with various timelines for enactment.

PROTECTED LAMPS IN METAL HALIDE FIXTURES

Metal halide (MH) lamps have the possibility of "non-passive failure" at end of life, which can cause hot quartz elements to exit the fixture.

The 2005 NEC addresses non-passive failure in Article 410.73(F)(5), "Metal Halide Lamp Containment," which states: "Luminaires (fixtures) that use a metal halide lamp other than a thick-glass parabolic reflector lamp (PAR) shall be provided with a containment barrier that encloses the lamp or shall be provided with a physical means that only allows the use of a lamp that is Type-O."

This means that either an enclosed fixture can be specified (with any type of lamp, including Type-E and Type-S), or an open-optic fixture can be specified that only operates Type-O lamps. These open fixtures feature a socket that can only operate Type-O lamps.

Type-O lamps are protected lamps, typical for 175-1500W sizes, that have additional containment around the arc tube. They feature a special base (EX39) so that they can only operate in compatible special sockets.

Fixtures with thick-glass PAR MH lamps are exempt.

The lighting industry has already been applying this rule to indoor open MH fixtures that operate lamps less than 350W. NEC is now applying it to all open MH fixtures. In the future, UL may review this requirement and if they adopt it as a standard for fixture manufacturers, the requirement will truly become national, since not all states may adopt the 2005 NEC quickly.

This requirement is not expected to significantly change specification practice. Specifiers will need to make sure that enclosed fixtures are suitably rated, and make sure open fixtures have a socket that only works with Type-O protected lamps. They will have to make sure that they specify Type-O protected lamps. Overall, this Code requirement is expected to simplify the MH systems options that are available.

METAL HALIDE AND MERCURY VAPOR FIXTURES IN SPORTS, MIXED-USE AND ALL-PURPOSE FACILITIES

In these facilities, the lamp's outer bulb can be broken during normal use of the space. When the bulb breaks, glass can fall out of the fixture into occupied space. In addition, the arc tube may continue operating, resulting in possible overexposure to UV radiation among occupants, which can cause sunburn and a burning sensation around the eyes.

The 2005 NEC addresses this situation in Article 410.4(E), which states: "Luminaires subject to physical damage, using a mercury vapor or metal halide lamp, installed in playing and spectator seating areas of indoor sports, mixed-use, or all-purpose facilities, shall be of the type that protects the lamp with a glass or plastic lens. Such luminaires shall be permitted to have an additional guard."

NEC requires that these lamps be completely enclosed with a glass or plastic lens to protect the lamp from damage. The fixture can contain an additional guard such as an external screen or cage, but this is not a substitute for the required enclosure.

DISCONNECTING MEANS DURING RE-BALLASTING

Industry data shows that a leading cause of fatalities among electricians is electrocution while working on 277V lighting systems. Some believe that this is partly because electricians are often pressured to change out ballasts while circuits are energized to avoid removing light from the area of servicing, causing them to ignore applicable warnings and instructions.

NEC has addressed this situation in Article 410.73(G), "Disconnecting Means," which addresses changes to how fluorescent fixtures are disconnected prior to electrical work to prevent the possibility of shock hazard. This Article states: "In indoor locations, other than dwellings and associated accessory structures, fluorescent luminaires that utilize double-ended lamps and contain ballast(s) that can be serviced in place or re-ballasted must have a disconnecting means, to disconnect simultaneously all conductors of the ballast, including the grounded (neutral) conductor if any. The disconnecting means must be accessible to qualified persons."

This requirement, however, will not become effective until January 1, 2008, to allow manufacturers time to comply. Basically, it appears to be a 2008 NEC requirement.

LIGHTING CONTROL PANELS

Industrial control panels used to control such systems as lighting, conveyor systems and air conditioning are, in many cases, manufactured in the field. The individual devices used in the system may be listed, but not the resulting panel itself. This has been a troubling issue for both installers and inspectors, as increased use of the panels has been accompanied by increased misapplication.

Specifically, in the event of an overcurrent situation, the energy level may exceed the short circuit current rating (SCCR) on a component in the system.

First, it would be useful to define what constitutes an "industrial control panel," since this term is not very commonly used in commercial applications.

NEC Article 409.2 states: "Definitions. Industrial Control Panel. An assembly of a systematic and standard arrangement of two or more components such as motor controllers, overload relays, fused disconnect switches, and circuit breakers and related control devices such as push-button stations, selector switches, timers, switches, control relays, and the like with associated wiring, terminal blocks, pilot lights, and similar components. The industrial control panel does not include the controlled equipment."

"The key to this definition is that the panel contains two or more of devices as stated in the NEC definition," says Scott Jordan, Marketing Manager for Square D/Schneider Electric. "As such, any enclosure containing a plurality of switching relays, an enclosure containing a relay and timer, or an enclosure containing a relay and a terminal block would fall under the definition as being classified as an industrial control panel.

"As such," he adds, "virtually all lighting control panels, supplied either by a manufacturer as a complete assembly or custom built on a job site, will need to meet the requirements of NEC 409."

The 2005 NEC addresses industrial control panels in a new section, Article 409, which is designed to facilitate safe installation as well as inspection. Previously, industrial control panels were installed based on general requirements from other NEC articles and special rules in some states. The new Article 409 covers general-use industrial control panels that operate at 600V or less. Article 409 impacts the way control panels are designed and built to ensure the entire panel and related components all meet a defined SCCR for the application, and that the panel is marked with the appropriate SCCR.

NEC Article 409.110 states: "An industrial control panel shall be marked with the following information that is plainly visible after installation: (3) short circuit rating of the industrial control panel based on one of the following: (a) short circuit current rating of a listed and labeled assembly; (b) short circuit current rating established using an approved method; FPN: UL 508A-2001 Supplement SB is an example of an approved method."

The SCCR provision in UL 508A is also new and becomes a national standard in April 2006.

NEC also requires: "SCCR for a component or equipment shall be equal to or greater than the available short-circuit current where the equipment is being installed in the system."

This NEC Article applies to OEMs, machine builders and panel builders, but it affects other professionals downstream as well. For example, if an existing panel is relocated, the state may require that 2005 NEC and its rules apply. If a panel is relocated after it is installed in compliance with 2005 NEC, the SCCR of the panel must be adequate for the new location. And inspectors will be looking for proper labeling on panels in new and updated installations.

Section IV

TECHNOLOGY

Chapter 15

Demand Reduction and Energy Savings Using Occupancy Sensors

*By the National Electrical Manufacturers Association,
Lighting Systems Division*

Lighting is one of the single largest users of electrical energy in a typical commercial building. While occupancy sensors have become a mainstream solution for eliminating wasted lighting energy in these applications, there continues to be a need for research documenting both the magnitude of the savings by application and the impact these controls have on demand. A study by the Environmental Protection Agency and the Lighting Research Center of Rensselaer Polytechnic Institute presented at the IESNA Annual Conference in Washington, DC (August 2000) provides unique and valuable data about occupancy sensor demand reduction and energy savings potential.

STUDY HIGHLIGHTS

Sixty organizations, which were active participants in the EPA's Green Lights Program, provided a total of 158 rooms falling into five occupancy types: 42 restrooms, 37 private offices, 35 classrooms, 33 conference rooms and 11 break rooms. Each room was monitored for occupancy and lighting status over a 14-day period using Watt Stopper's Intellitimer Pro light logger. The light logger data were converted to one-minute intervals, which made it possible to evaluate occupancy patterns, calculate energy savings and estimate the demand reduction potential using simulated occupancy sensor time delays. Occupancy sensor time delays of five, 10, 15 and 20 minutes simulated in the study, although data for the minimum (five-minute) and maximum (20-minute) time delay simulations are presented here.

ENERGY SAVINGS

The percentage of energy waste that actually occurred for the 14-day period and the calculated energy savings for the five- and 20-minute time delay simulations are summarized in Table 15-1. Not all of the wasted lighting energy is captured when occupancy sensors are used because lights remain on for the duration of the time delay setting. Similarly, the energy savings decreases as the timeout setting increases because lights remain on in the unoccupied room for a longer time period. Shorter time delays also increase the switching frequency of the lamps and ballasts, which may reduce lamp life.

Table 15-1. Energy waste for the 14-day period and energy savings for the five- and 20-minute time delay simulations.

Application	*Energy waste*[1]	*Energy savings using the 5-min time delay*[2]	*Energy savings using the 20-min time delay*[2]
Break Room	39 percent	29 percent	17 percent
Classroom	63 percent	58 percent	52 percent
Conference Room	57 percent	50 percent	39 percent
Private Office	45 percent	38 percent	28 percent
Restroom	68 percent	60 percent	47 percent

[1]Maniccia and Tweed, 2000
[2]Von Neida et. al., 2000

DEMAND REDUCTION

Demand reduction potential was analyzed by separating the analysis into a "daytime" analysis which analyzed the data from 6:00 am to 6:00 pm, and a "nighttime" analysis which analyzed the data from 6:00 pm to 6:00 am. Load profiles for each space type were also developed. The weekday load profiles for each space are illustrated in Figures 15-1 through 15-5. These graphs show the hourly time-of-day profiles for the actual energy use ("baseline"), and the load profiles for the actual energy use ("baseline"), and the load profiles from the 5- and

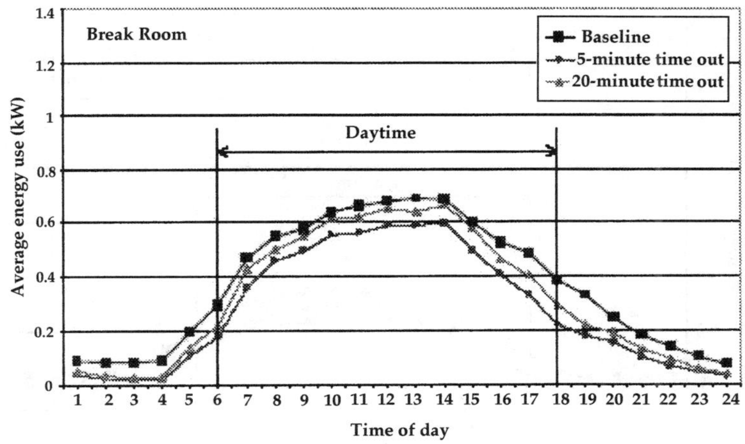

Figure 15-1. Break room.

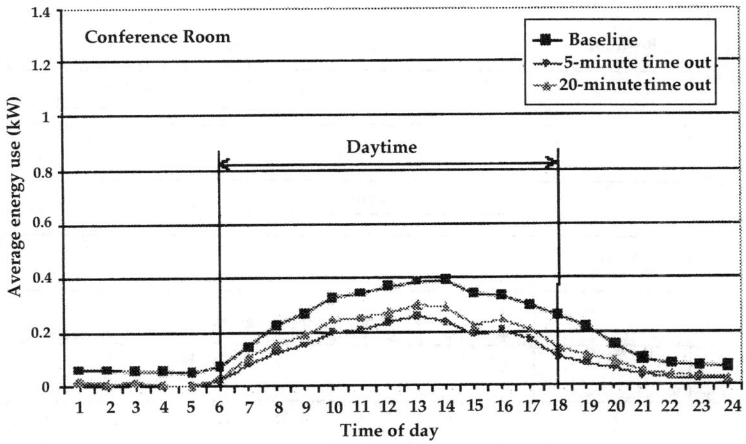

Figure 15-2. Conference Room.

20-minute time delay simulations. In all cases, the load profile is reduced when occupancy sensors are used.

The classroom data set includes both K-12 and higher education facilities data. The load profile for each of these segments for would like be different than the combined average shown here.

The average daytime energy demand reductions for the minimum and maximum time delay settings are listed in Table 15-2. These values represent the average reduction that occurs between the hours of 6:00

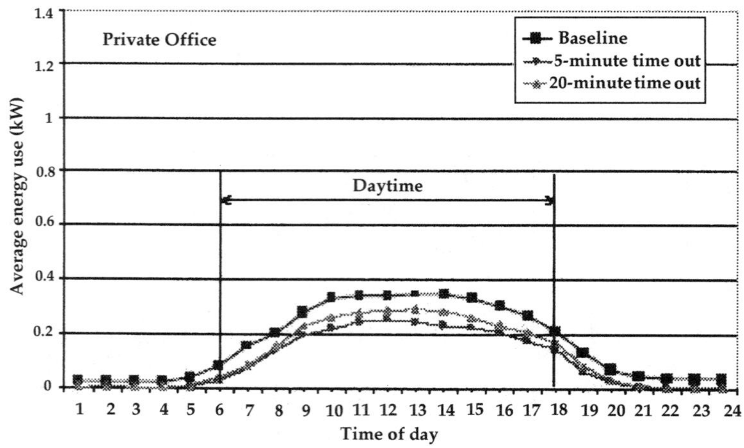

Figure 15-3. Private office.

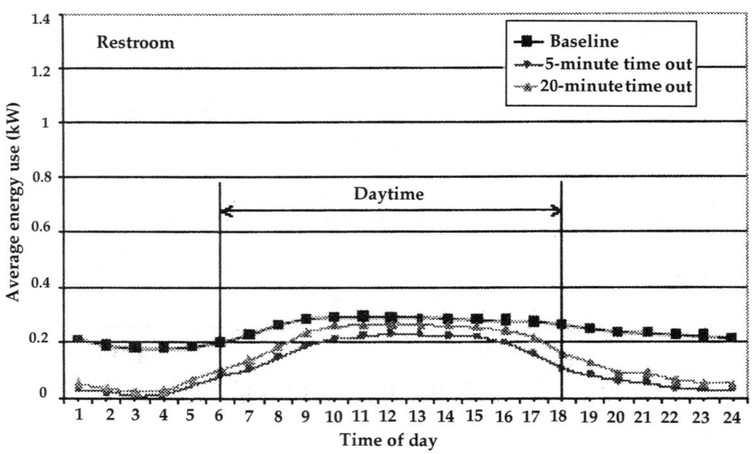

Figure 15-4. Restroom.

AM and 6:00 PM, and do not represent reductions at any specific time-of-day. An estimate of the magnitude of the reduction at a specific time of day can be garnered by comparing the baseline value from the graph to the value from the 5- or 20-minute timeout setting simulation.

Unlike changing out lamps and ballast to reduce the lighting watts per square foot, demand reduction with occupancy sensing reflects the fact that a portion of the individual spaces on a floor will be unoccupied at any point in time. The load profiles shown here illustrate that occu-

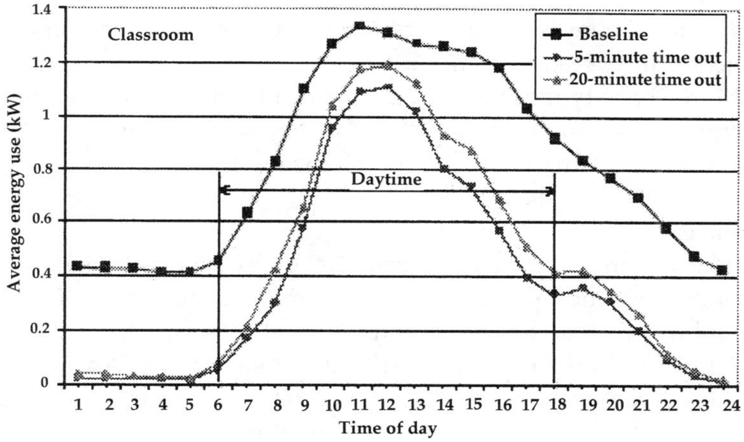

Figure 15-5. Classroom

Table 15-2. Weekday daytime average demand savings for the minimum and maximum time delay simulations[1].

Application	*Time delay*	*Daytime average energy demand savings[2]*
Break Room	5-min	18 percent
	20-min	8 percent
Classroom	5-min	40 percent
	20-min	31 percent
Conference Room	5-min	41 percent
	20-min	28 percent
Private Office	5-min	31 percent
	20-min	20 percent
Restroom	5-min	33 percent
	20-min	17 percent

[1]Von Neida et al., 2000
[2]Daytime demand savings are the average savings between 6:00 AM to 6:00 PM, and do not represent hourly demand reduction.

pancy sensors will reduce lighting energy use and demand throughout the day. The magnitude of the savings will depend upon the time delay setting and when the peak demand occurs, which may vary among building types. When looking at a large building with numerous individual spaces being controlled, the natural diversity factor will lead to a reduction in overall demand.

More space types need to be added to the test database, but it is clear from the results to date that occupancy sensors impact both total energy use and demand in individual enclosed spaces.

References

Maniccia, Dorene and Allan Tweed. 2000. Occupancy sensor simulations and energy analysis for commercial buildings. Troy, NY: Lighting Research Center, Rensselaer Polytechnic Institute.

Von Neida, Bill, Dorene Maniccia and Allan Tweed. 2000. An analysis of the energy and cost savings potential of occupancy sensors for commercial lighting systems. Illuminating Engineering Society of North America 2000 Annual Conference: Proceedings. New York: IESNA.

Chapter 16

Compatibility of Fluorescent Lamps and Electronic Ballasts in Frequently Switched Applications

By the National Electrical Manufacturers Association, Lighting Systems Division

Switching off fluorescent lamps whenever a room is unoccupied saves energy. New energy codes mandate automatic shut-off of non-residential buildings. Occupancy detectors, mandated switching off of lighting when leaving a room, and automated building systems save energy by removing lighting load when not being used. Frequent switching of lamps, however, may shorten their operating life. It is the intent of this paper to give some guidance in the selection of ballast type as a function of lamp switching rate to achieve the desired energy savings while maintaining acceptable lamp life.

SWITCHING FREQUENCY IMPACT ON LAMP LIFE

Frequent switching of lighting saves energy by removing the line voltage from lamp ballasts whenever the lights are switched off. When the power to the lamp ballast is restored, the lamp and ballast undergo a start. Lamps are rated to be started once every three hours during their life time. If lamps are switched more frequently than once every three hours, lamp-life will be reduced.

Studies have concluded that, even at significantly shortened life, the total life cycle economics may favor frequent switching, especially where energy rates are high (Louis Carrier and Mark S. Rea, "Economics of Switching Fluorescent Lamps," *IEEE Transactions on Industry Applications,* Vol. 24, No. 3, May-June 1988; see also *U.S. EPA Lighting*

Upgrade Manual, Fourth Edition, February 1993). If the operating time of lamps is reduced enough, *the chronological life* of the lamp may not be decreased at all through switching off by a frequent switching means. Similarly, if switching cycles are only moderated increased or if a ballast with a specially designed starting scheme is used, lamp life may not be affected. In either case, energy savings can normally more than offset lamp replacement costs.

GUIDELINES

The question becomes, how does one choose a ballast and switching scheme combination to minimize loss of lamp life in frequent starting applications?

To achieve acceptable lamp life, the specifier must address switching scenarios and ballast type.

"On-Time"

NEMA recommends that the minimum lighting "on time" be no less than 15 minutes. This allows for energy savings when people are out of the room for extended periods of time, but does not shorten lamp life by cycling lamps every time someone steps out of the room momentarily.

A product survey performed by the Lighting Research Center found that the vast majority of sensors would permit a minimum "on time" setting of 15 minutes and that many were adjustable to 20 and even 30 minutes. In the event that a given sensor is limited to less than 15 minutes, NEMA recommends setting the sensor to the longest time possible. If lamp life results at the 15-minute setting are unacceptable, then the time should be increased for those sensors with such flexibility. For the complete product survey, see "Specifier Reports—Occupancy Sensors: Motion-Sensing Devices for Lighting Controls," National Lighting Product Information Program, Vol. 5, No. 1, October 1998.

Ballast Type

There are three main types of ballasts, each with its own starting characteristic that can affect lamp life.

Instant start ballasts are the most efficient and the most popular electronic ballast available today. They are recommended for applications with switching frequencies of less than five cycles per day or where energy savings is considered more important than lamp life. Instant starting can make a ballast very efficient, but it causes the electrodes of the lamp to degrade a little every time the lamp lights compared with programmed start ballasts. An instant start ballast should start the lamp in the time specified by ANSI (ANSI C82.11-1993, *High Frequency Fluorescent Lamp Ballast,* and ANSI C82.11 Consolidated-2002, *High Frequency Fluorescent Lamp Ballast—Supplements*).

Rapid start ballasts are not as efficient as instant start ballasts due to additional filament heating power supplied to the lamp, although this additional filament heating can produce longer lamp life in applications where lamp starting occurs less often than every three hours. Like the instant start ballast, they are recommended for applications with switching frequencies of less than five cycles per day. Rapid starting of lamps causes the electrodes of the lamp to degrade a little every time the lamp lights compared with programmed start ballasts. A rapid start ballast should start the lamp in the time and manner specified by ANSI.

Programmed start ballasts provide the best lamp ignition and longest lamp life. In a programmed start ballast, electrodes are preheated prior to ignition, resulting in almost no electrode degradation. This allows frequent starts without a significant loss of lamp life. Programmed start ballasts are recommended in applications with frequent starts where extended lamp life is a primary concern.

SUMMARY: RECOMMENDATIONS

- Use the longest practical minimum "ON" time setting for the occupancy sensor and other automatic cycling means (15 minutes is recommended).
- Only use ballasts that meet ANSI requirements for lamp ignition.
- Use programmed start ballasts in areas that will result in a high number of switching cycles per day and where lamp life is a primary concern.
- Check with manufacturers for their recommendations on ballast/lamp/switching cycle compatibility.

References

The IESNA Lighting Handbook: Reference and Application, Mark S. Rea, Ninth Edition (New York: Publications Department, IESNA, 2000), 6-29 to 6-31. See Figure 6-38 for the effect of burning cycles on average lamp life for rapid start fluorescent lamps.

Specifier Reports—Occupancy Sensors: Motion-sensing Devices for Lighting Controls, National Lighting Product Information Program, Vol. 5, No. 1, October 1998.

U.S. EPA Lighting Upgrade Manual, Fourth Edition, EPA 430-R-93-001, February 1993. See the following figures: Lamp life versus cycle hours, fluorescent lamp life cycle cost, occupancy sensors and lamp life, and occupancy sensor economics.

Richard N. Thayer, "Determinants of Fluorescent Life," IES National Technical Conference, September 1954.

Louis A Carriere and Mark S. Rea, "Economics of Switching Fluorescent Lamps," *IEEE Transactions on Industry Applications,* Vol. 24, No. 3, May-June 1988.

Dorene Maniccia, Allan Tweed, Bill Von Neida and Andrew Bierman, "The Effects of Changing Occupancy Sensor Timeout Setting on Energy Savings, Lamp Cycling, and Maintenance Costs," (Troy, NY: Lighting Research Center, August 16, 2000). See Figure 1 on expected lamp life for operating cycles shorter than three hours per start for instant start systems.

Ballast-Lamp Technology Update, "Fluorescent Lamp Starting and Operating Technologies" (Danvers, MA: Sylvania, July 17, 2000). See the figure pertaining to minutes per switch cycle versus average life.

Chapter 17

Digital Lighting Networks Offer High Energy Savings and Unprecedented Flexibility in Lighting Control

By Craig DiLouie, Lighting Controls Association

Imagine a large office building where the lighting system is set up as a dynamic network with each fixture able to be controlled separately or in multiple combinations of groups—then reprogrammed as space needs change. The fixtures can be both locally and centrally dimmed or switched, and continuously provide energy information to a central computer that can be used to identify lamp and ballast failure, generate load profiles and verify energy savings.

This unprecedented level of lighting system control is now possible using an open lighting networking scheme based on products compatible with the Digital Addressable Lighting Interface (DALI) protocol. With DALI, dimming—traditionally restricted to architectural lighting, daylight harvesting and conference rooms—can be deployed across the facility to achieve high energy savings (30-60 percent including ballast-lamp efficiency) as well as extraordinary flexibility, user control and maintenance benefits.

DALI, part of IEC Standard 929, provides communication rules for lighting components, first developed in the mid '90s, with commercial application begun in 1998. In Europe, DALI has been adopted as a new standard by ballast manufacturers including Osram, Philips, Tridonic, Trilux, Helvar, Hüco and Vossloh-Schwabe. DALI is now making an entry to the U.S. and is gaining interest from manufacturers in building DALI-compatible ballasts and controls, some of which are now coming out on the market.

Digital addressable ballast capabilities are changing the way that the industry designs and controls space. DALI provides a vehicle for manufacturers, building managers and lighting management companies to have confidence that products from multiple manufacturers will be compatible and interoperable.

HOW IT WORKS

While the system sounds sophisticated, most of the hardware is common—ballasts, lamps, controls, wiring—with the difference that the ballasts are connected to a central computer enabling each ballast to be individually addressed, programmed and grouped.

A DALI-based digital lighting network is based on digital 120/277V fluorescent electronic ballasts, currently available in one- and two-lamp models that operate linear T5, T5HO and T8 fluorescent lamps as well as compact fluorescent lamps. According to Tridonic, digital ballasts and DALI interfaces will soon be available for high-intensity discharge (HID), incandescent and low-voltage halogen systems. Digital ballasts "soft start" fluorescent lamps to increase service life; cut the lamps out at end of life; gradually dim; and start the lamps at any point in their dimming range, from 100 percent to 1 percent.

The ballasts are connected using either Class I line-voltage or Class II low-voltage wiring to form a lighting bus or loop of up to 64 ballasts. Each ballast is given an address in the system so that it can be individually controlled or grouped in multiple configurations (up to 16 layers of control/scenes). The loop is then connected to any type of DALI-compatible control device(s). Control options include local wall-mounted controls that enable manual push-button switching to select programmed dimming scenes; a computer for centralized lighting control; local PCs for individual occupant control; and occupancy sensors, photosensors and other controls.

As a digital lighting network is relatively sophisticated, it is generally suited for large installations and, as in the case of an energy management system, requires planning and time to program various instructions into the computer. Its capabilities are ideally suited for small and open offices where users can control their own lighting; conference rooms and classrooms that require different lighting scenes for multiple types of use; supermarkets and certain retail spaces where merchandising and layout changes frequently; hotel lobbies and meeting spaces to accommodate times of day, events and functions; and restaurants to match the lighting to time of day (breakfast to lunch to

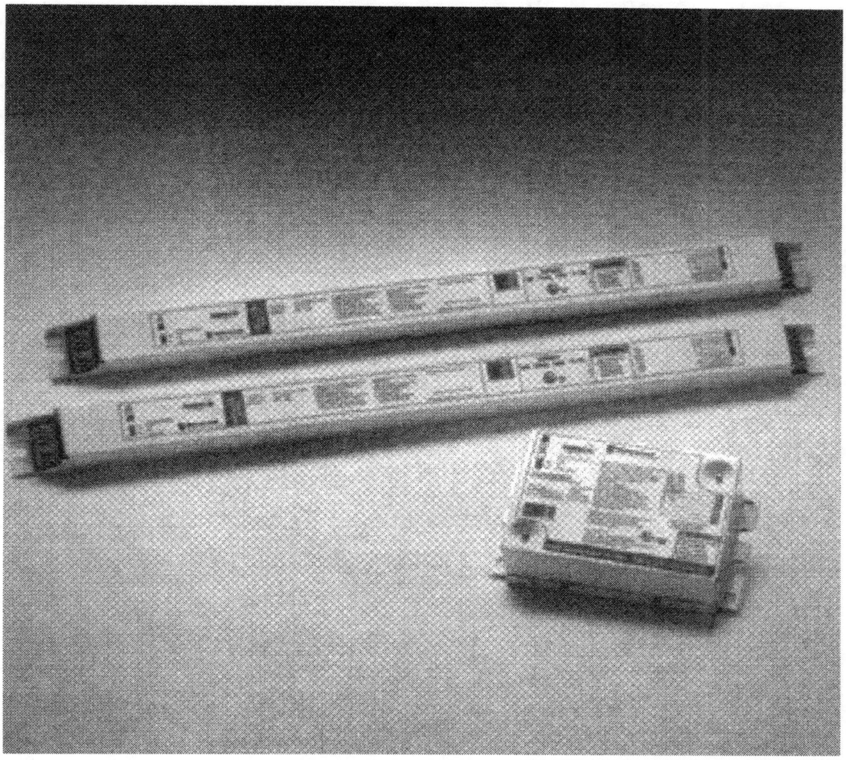

Figure 17-1. Digital ballasts. *Source: Universal Lighting Technologies*

dinner).

Digital lighting networks offer substantial benefits but as with any new technology, it is still coming of age. Below, we will examine the benefits and the progress the lighting industry has made in developing products that capitalize on the DALI protocol.

BENEFITS

The most significant benefits of a digital lighting network are its high energy savings, flexibility and maintenance potential. Customized dimming across the facility, with a level of occupant control, can be used to make each fixture responsive to both prevailing conditions (peak demand, energy rates, available daylight, occupancy, type of task)

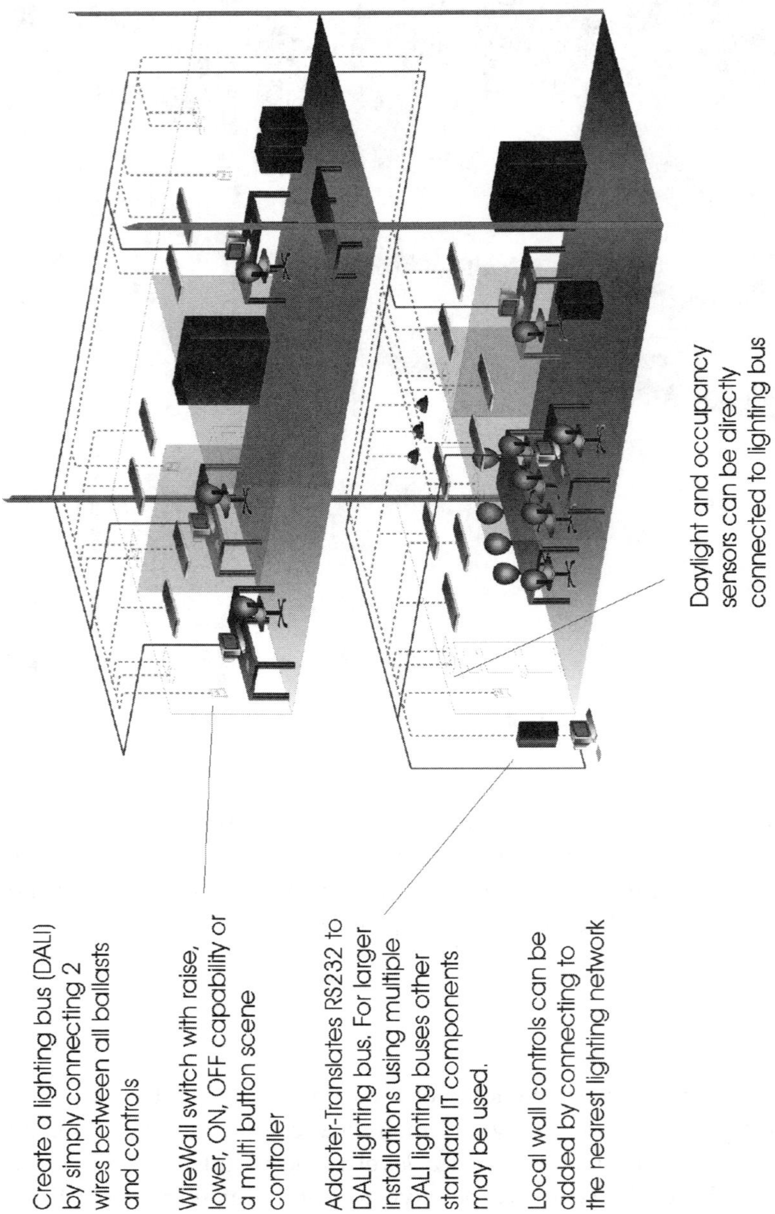

Figure 17-2. Application of a DALI-based system in an office building. *Source: Tridonic*

and local occupant needs. The energy savings potential of daylight harvesting, peak-demand dimming and occupancy sensors is well-documented. Studies conducted by the Lighting Research Center demonstrate that users enjoy personal control of their light levels. And significantly, DALI-based lighting control systems submeter (to a degree) fixture performance, enabling new maintenance and energy information possibilities.

Energy Savings

The lighting network enables each individual digital ballast to be controlled by a system that includes a static element (programmed dimming or on/off based on time of day) and a kinetic element (switching or dimming in response to sensed occupancy or ambient light level). By taking advantage of the inherent efficiency of digital electronic ballasts working in tandem with various efficient light sources and dimming and switching strategies (occupancy sensors, scheduled on/off switching, daylight harvesting, lumen maintenance dimming), energy consumption can be reduced by 30-60 percent. Energy savings can be accelerated by establishing dimming setpoints for various loads during peak demand periods to reduce utility demand charges.

In addition, each ballast feeds performance data back to the central computer, including energy information and signals of lamp or ballast failure (see maintenance benefits below). When dealing with large fixture groups, the energy information produced by the lighting network is closely approximate, which can be useful to verify energy savings, generate load profiles, support internal billing and produce a lighting cost per unit of production. Submetering may be desirable, however, if exact information is required, as is usually the case in fulfilling a performance contract between a building owner and an ESCO.

Flexibility/Productivity

The digital lighting network provides an open environment in which any combination of ballasts can be grouped and controlled in multiple ways depending on prevailing task needs, occupant preferences and changes to primary tasks with the space. The control system can be configured so that individual occupants can dim or increase the light output of the fixtures serving their workspace, enabling occupant-driven fine-tuning of light levels based on tasks, worker age, etc. In addition, when the space is remodeled or if its task needs change, the fixtures do not have to be moved and rewired; instead, in many cases, the owner can simply reconfigure and program the ballasts to provide the optimum lighting conditions.

Maintenance

As noted above, fixtures in a digital lighting network continuously provide energy information to a central computer. This enables monitoring of anomalies across the entire lighting system, which provides alerts that immediate replacement is needed of lamps and ballasts or that troubleshooting may be required. The software indicates what type of component needs to be replaced (lamp or ballast), what type of lamp it is (for example, T5, T5HO, T8) and where the fixture is located (for example, Building 3, 4rd Floor, Office A-12, fixture over desk #3).

Coming of Age

Digital lighting networks are in the late introduction phase in the U.S. market. Early adopters are currently using the technology, which will certainly be used by manufacturers to validate the concept to the marketplace. And a broad range of manufacturers are currently developing products that support the DALI protocol. The benefits are real; the concept is being validated; but it may take a little more time to have access to the full range of control product options and competitive ballasts. As demand grows, the ballast and controls industries will be ready to respond.

Chapter 18

BACnet: Introduction to the Building Automation Standard Protocol

By Craig DiLouie, Lighting Controls Association

While the Digital Addressable Lighting Interface (DALI) protocol has generated a lot of buzz in the lighting industry in recent years for its extraordinary benefits, a standard protocol introduced by the American Society of Heating, Refrigerating and Air-Conditioning Engineers (ASHRAE) in 1996 continues to have similar strong implications for lighting. It's called BACnet, a data communication protocol for building automation and control networks. It is an American (ANSI) standard and an ISO global standard.

Like DALI, BACnet is not a product. Instead, it provides a set of rules that govern the exchange of information over a computer network. BACnet-compliant lighting, HVAC, security and other systems and devices from multiple vendors can be attached to this network for ideally single-seat workstation global control over all connected building control systems.

Although one of the benefits of BACnet is that it's scalable from small to large installations, the protocol is following the market for building automation systems, which are being purchased for larger installations. Building automation systems are typically associated with buildings 100,000 sq.ft. and up. A good example might be a university campus, with different buildings that use equipment from different manufacturers, all of which must be networked together with a single point of control. BACnet solutions are often seized upon because they solve a problem that requires interoperability.

BENEFITS OF INTEROPERABILITY

BACnet is the result of decades of interest and effort in the HVAC industry. What makes it important to lighting is that specifiers and owners are showing high interest in tying all building control systems into a single point of control.

According to a market research study funded by The Watt Stopper and conducted by Ducker Research, now available as part of the California Energy Commission's Public Interest Energy Research (PIER) Lighting Research Program, specifiers and owners of lighting automation systems want the benefits of interoperability. Of five trends influencing the controls field, standard protocols, and ability to integrate the lighting automation system with the building management system, ranked as most important.

Below are the primary benefits of interoperability:

Lower Costs

Most engineers and other specifiers prefer to work with a small group of vendors or even a single vendor, but desire the economy of competitive pricing during bidding. If all systems and devices are interoperable, specifiers can mix and match and select products in a competitive environment to create the best solution at the lowest cost.

Single Point of Control

Users are interested in establishing a single point of control over all building systems, from HVAC to lighting, with an operator at a consolidated point of control. This arrangement greatly simplifies building management and creates synergies among various building control systems to save energy and perform profiling. For example, the occupancy sensors that connect to a BACnet lighting management system can also be used to signal the HVAC system so that it can switch between "occupied" and "unoccupied" setpoints.

Confidence in System Performance

An industry standard provides a level of assurance that various compliant devices from different vendors work together in a system.

Flexibility and endless useful life for systems: Facility use and automation capabilities change quickly, which can render existing systems obsolete if new product innovations are not available from the original vendor and are therefore not interoperable with the existing system. If all products are plug and play on a network, then new innovations can be easily integrated, which can help ensure perpetual useful life of the system.

BACNET SPECIFICATION ISSUES

Product Testing

BACnet-compliant systems and devices are tested by the National Institute of Standards and Technology's (NIST) BACnet Interoperability Testing Consortium. In 2000, the BACnet Manufacturers Association founded the BACnet Testing Laboratories (BTL), and began testing products for compliance with the BACnet standard in December 2001. Its testing procedures are based on the BACnet testing standard approved by ASHRAE in 2003.

Product Availability

According to NIST members, there were currently more than 4,000 BACnet sites operating in the United States and other countries in 1998, of which about one-third are multivendor installations. By 2000, more than 19,000 installations were found to be in operation, according to a study conducted by the BACnet Manufacturers Association. A number of manufacturers have committed to BACnet. More than 120 manufacturers have received BACnet vendor ID numbers. A list of a portion of these companies can be seen at www.bacnet.org.

Challenges

While BACnet is a standard protocol, it faces competition from LonTalk, an open protocol offered by Echelon as part of the company's LonWorks network technology. LonTalk is a LAN specification, and can be used to communicate BACnet control messages. Echelon, however, has its own control language that also uses LonTalk and although both can protocols can use a LonTalk LAN, for controls communication they are now in competition. Some users say that BACnet will prevail, while others say LonTalk will, and others say that the two each have their niche and can exist side by side. The real problem for users stems from the fact that manufacturers, for the most part, have aligned themselves with one or the other of the two protocols, which limits product availability.

In addition, all protocols face the same problem: Specifiers are slow to use it if products are not available, and manufacturers are slow to develop products if specifiers aren't requesting it. Supply and demand seem to inch forward together, a process that takes time. However, the rapid increase in BACnet installations since the protocol's introduction appears to be a promising indicator of market penetration.

Specification Issues

The performance of a BACnet product is captured in its protocol implementation conformance statement (PICS) spec sheet. The PICS sheet provides a list of the product's BACnet capabilities, such as what LAN options are available.

The PICS sheet is very valuable during specification, which is sort of a PICS in reverse in which the specifier writes down what network functions are needed—such as alarm and event requirements, points shared between devices, etc.—and states that these functions must be provided using BACnet.

Networking Issues

BACnet systems are connected using a number of networking options, as shown in Table 18-1.

The advantage of all of these options is that BACnet messages can be conveyed by virtually any network technology, whichever is required or most cost-effective. The downside is that since BACnet uses several different architectures, BACnet-compliant devices may still have interoperability problems on the same set of wires. This risk of this is fairly low, however, and can be avoided altogether through the use of the PIC statement from each manufacturer.

If a given site includes more than one type of network, a router that follows the BACnet standard can be specified. A router is simply a device used to transfer messages from one network to another.

If a given site includes a legacy network to be connected to a BACnet network, a gateway is required. A gateway is different from a router in that it doesn't simply transfer messages; it also translates them into each network's local language. This device can also be used to exchange messages with a LonWorks network. BACnet gateways are special items and can add significant cost and complexity to a project. They also present a single point of failure.

Table 18-1. Networking options for connecting BACnet systems.

Network Technology	*Speed*	*Physical Media*
Ethernet	10 Mbps (100 now available)	Coaxial cable, twisted pair, fiber-optic, wireless, etc.
BACnet/IP		
ARCNET	2.5 Mbps	Coaxial cable, twisted pair, -optic, etc.
Point-to-Point (PTP)	115.2 kbps	Serial cable, phone lines
LonTalk	2.5 Mbps	
MS/TP	1 Mbps or less	Twisted pair, wireless

BACNET AND LIGHTING

Specifiers and owners can gain the benefits of interoperability by either choosing a native BACnet lighting management system or a DALI lighting network. If a DALI network is chosen, it can be integrated with a BACnet lighting management or building management system using a gateway (see Figure 18-1, following page).

A Lighting Applications Group was formed in January 2001 so that representatives of lighting control manufacturers and other industry groups can address and solve lighting-specific issues and applications in the BACnet standard.

In the lighting industry, a number of BACnet products are available from manufacturers such as Touch-Plate, Musco Lighting, The Watt Stopper and Lithonia Lighting.

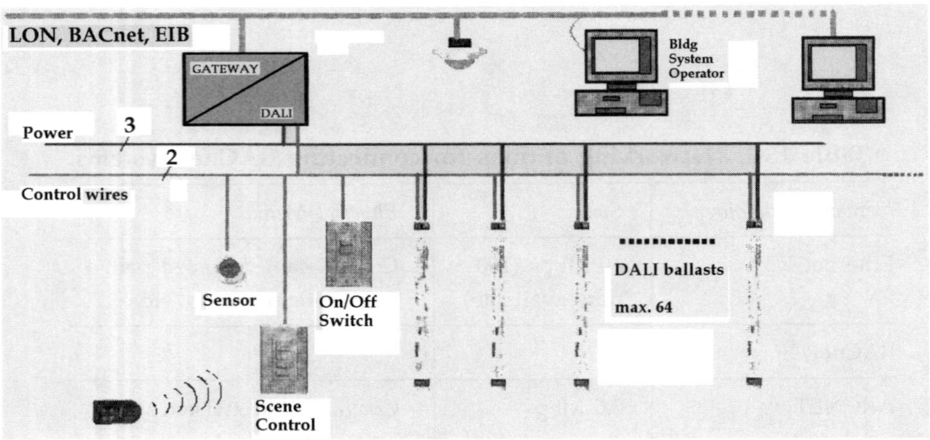

Figure 18-1. A DALI network can also be integrated into a building management system by adding a suitable gateway to "translate" back and forth between DALI and BACnet, Lonworks or other systems. (Consult with your building management system supplier for specifications and availability.) (Note that the diagram is simplified; the BMS/DALI gateway block diagram also serves as the DALI loop power supply; the two functions were combined in one block to save space.) Illustration courtesy of OSRAM SYLVANIA, Inc.

Chapter 19

Linear Fluorescent Dimming Ballasts: Explaining the Protocols

By Craig DiLouie, Lighting Controls Association

Dimming linear fluorescent lamps can provide a number of significant benefits to owners of commercial lighting systems:

- Flexibility, enabling the lighting system to adapt to multiple activities and changing space needs.
- Cost savings, derived from direct energy savings as well as load reduction during peak demand periods, which can be accelerated by using dimming ballasts in a system that can also include occupancy sensors, daylight sensors and time-clocks.
- Higher worker comfort, satisfaction and performance, achieved by allowing occupants to choose their own light levels.
- Increased lamp life for applications where lamps can be dimmed instead of frequently switched.

Dimmable fluorescent systems combine the long life and energy efficiency of fluorescent lamps with the controllability and full-range dimming capabilities of incandescent systems. In this white paper, we will discuss how linear fluorescent lamps are dimmed, then compare popular methods for dimming with pros and cons of each.

DIMMING BALLASTS

Linear fluorescent lamps produce light when an arc of electric current is established across the lamp from one cathode to the other, causing the gas to emit energy that is converted into visible light by the phosphor coating the inside of the glass bulb. Fluorescent lamps require a ballast to operate, an electrical device that provides the proper starting voltage to initiate the arc and then regulates the current flowing through the lamp.

The ballast can be configured so that it 1) receives a signal from a control device and subsequently 2) changes the current flowing through the lamp, thereby achieving a gradual controlled reduction in lamp output. The characteristics of the control signal affect the duration and extent of the change in current and subsequent lamp output.

Dimming ballasts are available for operation of linear and compact fluorescent lamps. This chapter focuses on linear fluorescent lamps.

Most commercially available dimming ballasts for operation of these lamps are electronic rapid-start or programmed-start ballasts, and all linear lamps operated by these ballasts feature bi-pin bases typical of rapid-start lamps.

Rapid-start ballasts preheat the cathodes with a small voltage, which reduces the amount of voltage needed to start the lamp. After preheating the cathodes, the ballast provides the high voltage required to initiate the arc.

Programmed-start ballasts are rapid-start ballasts that preheat the electrodes more accurately to minimize damage to the electrodes during the start-up process (according to a program) and therefore can optimize lamp life. While supplying the preheat voltage, the ballast minimizes the lamp voltage, thereby reducing glow current during this phase with its associated degrading effect on lamp life. As a result, programmed-start ballasts can provide up to 100,000 starts, ideal for applications where the lamps are frequently switched, such as space with occupancy sensors.

DIMMING METHODS: ANALOG VS. DIGITAL

Several methods can be used to achieve the dimming effect. Because the dimming ballast must be able to communicate with connected controllers, the method becomes the basis for a protocol, or common operating parameters adopted by all manufacturers of dimming ballasts and controllers that use that method. This assures interchangeability between the ballast made by a particular manufacturer and various controllers made by controls manufacturers. Check the ballast manufacturer for compatibility between its ballasts and various controls.

The primary methods are:

Analog

The analog electronic dimming ballast includes components that perform these functions: electromagnetic interference filtering, rectification, power factor correction and ballast output to power the lamp. There are several analog methods, including 0-10VDC, two-wire phase-control, three-wire phase-control and wireless infrared, with 0-10VDC being most popularly used.

Digital

The digital electronic dimming ballast includes components that perform these functions: electromagnetic interference filtering, rectification, power factor correction, a micro-controller and ballast output to power the lamp. The micro-controller functions as a storage, receiver and sender of digital information. The micro-controller can store the ballast address, receive control signals and send status information.

Analysis

Analog dimming systems are established and common, while digital dimming systems are relatively new to the industry. Both provide the essential function of controlling the lamp output based on input from a control device. Both enable the construction of networks of controls and ballasts wired to local and central points where control signals can originate, either manually or based on a program.

Analog is the standard dimming method, typically presents a lower cost, and is compatible with a wide range of common control devices. The dimming ballasts can be on a low-voltage or line-voltage control circuit. Analog ballasts and controls are compatible as long as they are configured to the same method—e.g., 0-10VDC, etc.

Digital provides a higher degree of granularity of control capability, such as ability to individually address and group the ballasts, gain feedback information from ballasts, manage a variety of zones and scenes, and provide a lighting system that can easily accommodate changes over time.

PROTOCOLS

Dimming ballasts must be configured to understand and act upon the control signal coming from a control device over either low- or line-voltage wires. To ensure compatibility, protocols have been developed around the various dimming methods.

It should go without saying that items designed to operate on different protocols are not compatible and should not be operated together. Doing so will result in the dimming system failing its purpose, as well as potentially damaging the equipment.

Analog

There is currently no standard for the operation of analog dimming ballasts. While there is a 0-10VDC control ANSI standard for the entertainment industry, it does not apply to dimming ballasts. As a result, equipment may work well together as a system but dimming performance may not be consistent among different ballast types and ballasts made by different manufacturers. A 5V signal for one ballast might result in a 50 percent dimming level but 30 percent on another, for example.

Digital

For digital ballasts, the Digital Addressable Lighting Interface (DALI) protocol, part of Europe's IEC Standard 60929, provides a standard. DALI offers the possibility of true interchangeability between ballast manufacturers and defines light output for all levels of dimming signals, ensuring consistent dimming performance across all dimming ballasts regardless of type or manufacturer. This ensures that different ballast types can mingle in the same control area and simplifies commissioning.

METHODS/INTERFACES

The dimming method is an important consideration, since it often defines the range of possible change in the lamp output and also the wiring configuration, which in turn affects capability as well as cost. As with everything in lighting, there are trade-offs.

Digital

Digital ballasts are recommended to use a Class 1-rated 5-conductor cable that uses one hot (live), one neutral, one ground and two polarity-insensitive control wires, all routed together in the same conduit. It is also possible to install the ballasts and controls as a Class 2 installation, in which case the control wires must be routed through separate conduit as the power wires. Check with the ballast and controls manufacturers whether their products are rated for Class 1 installation.

Manufacturers of DALI-based digital ballasts include Advance Transformer Co., Lutron Electronics, OSRAM SYLVANIA, Tridonic USA and Universal Lighting Technologies.

The other digital protocol is proprietary, developed by Energy Savings Inc. (ESI), which was purchased by Universal, whose digital ballasts are now marketed under the AddressPro brand.

Analog (0-10VDC)

Dimming is accomplished by controlling the amplitude of the current flowing through the lamp via reduction in the lamp power. As lamp power decreases, lamp voltage increases proportionally to maintain heating of the lamp cathodes and prevent the lamp from being extinguished.

0-10VDC ballasts use four wires, with two line-voltage leads (hot and neutral) to power the ballast and two low-voltage control leads to change the light level. Depending on wire insulation and control switch ratings, the control wires may either be routed in the same raceway (Class 1) or in a separate raceway (Class 2). In general, the system may be installed as Class 1 if the control wires carry the same voltage rating as the power wires and the control device is rated for Class 1.

This wiring scheme adds labor and material costs to the installed system cost, but enables the dimming ballast to be linked to other ballasts and control devices in a larger system, which in turn can be linked to local occupant controls and central control.

Typically, 0-10VDC ballasts have violet and gray control wires. The gray wire is internally connected to provide a ground reference. When the voltage level is near or above 10VDC, the ballast responds with full light output. As the voltage decreases, the ballast reduces light output. The ballast can also be connected to a switch or relay to enact bi-level control, providing full light output when the switch opens and reducing it to a specified minimum when the switch closes.

Note that some manufacturers provide command regions in the 0-10VDC range; a signal less than 0.3V might signal the ballast to shut down, for example. Be sure that the specified controllers are compatible with any such added feature for the chosen ballast.

Manufacturers of 0-10VDC dimming ballasts include Advance

(Mark VII), Lutron (TVE), OSRAM SYLVANIA (Quicktronic Helios and PHO-DIM), Universal (Ballastar, SuperDim) and GE.

Analog (Two-wire Phase-control)

Also called AC dimming, phase chop dimming or two-wire dimming, phase-control dimming entails "reading" the AC power supply signal's "starting point" or zero crossing point, then turning on the current after a preset waiting time. This "cuts out" part of the cycle and results in dimming. The extent of the waiting time, usually 0-8.3 milliseconds or one-half the waveform, is related to the dimming level.

Phase-control ballasts use the same two line-voltage leads for both power and ballast control. The ballast receives the dimming signal through the dimmed hot wire connected to the power line.

Because the standard wiring configuration is utilized, phase-control dimming ballasts represent a lower-cost dimming solution, typically found in architectural dimming applications such as conference rooms, boardrooms and individual offices. It is also ideally suited to retrofits, stand-alone applications and cost-sensitive projects. In addition, the control signals are less sensitive to interference than low-voltage analog signals.

At the time of writing, Advance is the only manufacturer that offers a full line of phase-control dimming ballasts (Mark X Powerline). Lutron makes available a limited offering (Tu-Wire). OSRAM SYLVANIA and Universal Lighting Technologies have discussed developing such ballasts and offering them in the near future.

Analog (Three-wire Phase Control)

The three-wire phase control configuration is based on the original magnetic dimming ballast conventions from the 1960s. This control method uses a third wire (in addition to hot and neutral) to carry the (typically) phase control signal to the ballast. All three wires are rated Class 1 and can be run within the same conduit. At the time of writing, Lutron manufactures three-wire phase-control dimming ballasts (Hi-lume, Compact SE and ECO-10).

Wireless Infrared Control

Some manufacturers also have wireless infrared control available. This method uses an IR transmitter to perform the control function and does not require any additional wires. The dimmer is included either in

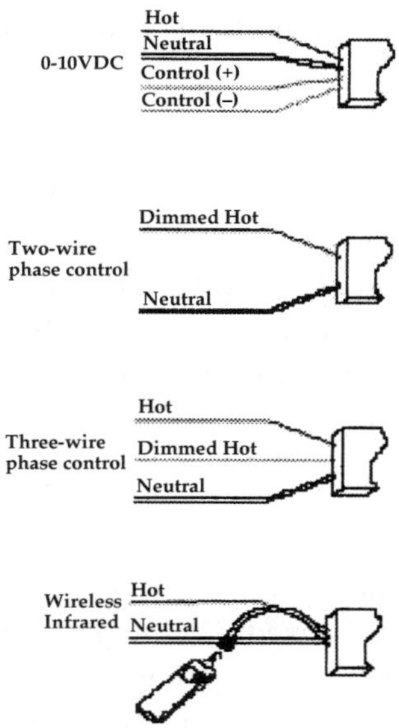

Figure 19-1. Analog dimming methods. *Courtesy: Lutron Electronics*

the ballast or as an additional device in the light fixture. This may be considered a good retrofit solution, and allows for occupant fixture control. At the time of writing, these types of ballasts are available from Lutron (ECO-10).

FINDING THE BEST SOLUTION (AMONG ANALOG BALLASTS)

A major difference between the three main analog dimming ballasts is the equipment required to control them. They are all "hard-wired" to the control circuit or zone, and one control device can control one zone. All of the ballasts wired to the same purple and gray wires (0-10VDC) and wired to phase-cut dimmed leg for two- and three-wire control will be controlled together. For building-wide control, these control wires must be connected to some type of dimmer which is then connected to the other dimmers and some type of building-wide network, presumable with some type of central control.

There are two main system topologies for this system:

Centralized

All dimming control wires for an area are pulled back to a dimmer cabinet or cabinets mounted in the electrical closets, and then these dimmer cabinets are connected together and to a central controller via a network.

Distributed

The dimming control wires are connected to a device that is mounted nearby, such as on the wall or in the plenum, and then these control devices are all connected together and to a central controller via a network. Either topology can be used to achieve building-wide control.

The various analog ballast types have different advantages in these systems:

0-10VDC

0-10VDC ballasts have the advantage of needing only small low-voltage components in the control device, so they are easiest to use in a "distributed" system. 0-10VDC control allows the on/off control to be separated from the dimming control, allowing a combination of centralized switching and distributed dimming equipment to be used.

Two-wire
Phase-control

Two-wire phase-control ballasts have the advantage of not needing any additional wiring between the control device and the ballast, which makes them very attractive for new centralized dimming applications as well as retrofits. They also don't require a separate switched power leg, so the hardware required to dim these ballasts is exactly the same as the hardware required to dim incandescent loads. This means that most (if not all) dimmer manufacturers include a way to adjust the dimming curve of their dimmers to allow the control of two-wire ballasts from their dimmer cabinets.

Three-wire
Phase-control

Three-wire phase-control ballasts draw very little current on the dimmed leg, which means that they can be dimmed without causing much heat to be generated at the dimmer. This allows devices that are intended only for this type of load to be smaller and also appropriate for use in a distributed system.

While considering all of the factors, the best solution for any given application, of course, depends on the application need. For example, is the primary goal energy savings, visual need or some other application need? What kind of dimming performance is required—100 percent to 1 percent, 5 percent or 10 percent? The choice of dimming ballast often comes down to specifier preference, dimming system compatibility, total installed cost (including wiring), and availability for the fixtures being used.

Table 19-1. Comparison of Dimming Protocols.

Comparison between ballast control methods				
Digital	*0-10V*	*Two-wire phase Control*	*Three wire phase control*	*Infrared control*
Dimming range: 1%-100% dimming ballasts are available.	Dimming range: 3%-100% ballasts are available for T8 lamps; 1%-100% ballasts are available for T5HO lamps.	Dimming range: 5%-100% available for T8 lamps; 1%-100% available for T5HO lamps	Dimming range: 1%-100% available.	Dimming range: 1%-100% available.
Wiring configuration: It is recommended that a five wire Class 1 rated cable is used. The ballasts and control devices must be Class 1 rated. Otherwise, the control wires have to be routed separately from the power wires.	Wiring configuration: Two power wires are run through the conduit carrying line voltage wires. The control wires are Class 2 and are not allowed in the same conduit. Some local codes require a separate Class 2 conduit.	Wiring configuration: Both power and control are routed through the same line-voltage wires. This ballast wires the same way as a conventional non-dim ballast.	Wiring configuration: All wires are Class 1, and relative to the phase control ballast, there is an additional control wire which is routed in the same conduit as the other wires.	Wiring configuration: No additional wires are required outside the fixture. The dimming device is either integral to the ballast or a separate interface within the fixture.
Typical applications: Small and open offices where users can control their own lighting; conference rooms and	Typical applications: Ideally suited for energy management systems. New construction and retrofit installations:	Typical applications: While two-wire ballasts can be incorporated into building-wide control systems, according to their primary manufacturer they are ideally suited for	Typical applications: Ideally suited for architectural dimming. Conference rooms, boardrooms, patient/examination/treatment rooms, houses or worship, theaters, convention areas,	Typical applications: Ideally suited for spaces where individual control is desired without additional wiring. Conference

Table 19-1. (Continued

Comparison between ballast control methods

Digital	*0-10V*	*Two-wire phase Control*	*Three wire phase control*	*Infrared control*
classrooms that require different lighting scenes for multiple types of use; supermarkets and certain retail spaces where merchandising and layout changes frequently.	auditoriums and training areas, conference rooms and boardrooms, department and specialty stores, education, healthcare, hotels, houses of worship, private and executive offices, restaurants.	architectural dimming, stand-alone, retrofit and low-cost projects. New construction and retrofit installations: auditoriums and training areas, conference rooms and boardrooms, department and specialty stores, education, healthcare, hotels, houses of worship, private and executive offices, restaurants.	restaurants, air traffic control centers, industrial control rooms, graphic art workstations, CAD/CAM workstations, private offices	rooms, board rooms, open and private offices.
Controlled by: Building automation system or lighting automation system. Occupant override through PC or local preset controller.	Controlled by: Energy management systems and occupants.	Controlled by: Local controls accessible to the occupants.	Controlled by: Central control systems and local controls accessible to the occupants	Controlled by: Individual controls (infrared transmitters) given
Available from Advance, Lutron, OSRAM SYLVANIA, Tridonic, Universal.	Available from Advance, Lutron, OSRAM SYLVANIA, Tridonic, Universal.	Available from Advance, Lutron.	Available from Lutron.	Available from Lutron.

Table 19-1. (Continued

Comparison between ballast control methods

Digital	*0-10V*	*Two-wire phase Control*	*Three wire phase control*	*Infrared control*
Bottom line: Installed component cost can be higher than comparable 0-10VDC systems due to power supply/router requirements, but the total installed cost can be installed cost after considering the wiring labor for group and scene control. Flexible system that offers individual ballast control and status feedback. Allows software configuration of lighting groups, presets matching the lighting to the space usage, and integrated energy management functions. May be configured as a large networked networked system requiring requiring commissioning and and training or as simple stand-alone room preset dimming controls requiring no special tools or PCs. Components of different manufacturers can be combined in the same installation.	Bottom line: Energy savings through building management system and occupant control.	Bottom line: Architectural dimming system, ideal for conference rooms, etc. as well as stand-alone and retrofits, and can be integrated into a buildingwide system.	Bottom line: Individual control system which can also be integrated into a buildingwide control system.	Bottom line: Architectural dimming system that can be integrated into a buildingwide system.

DIMMING ISSUES

Important issues related to dimming include perceived brightness, perception of light level reduction, power quality and energy consumption.

Perceived Brightness

As lamps are dimmed, light level decreases but the human eye may perceive a higher light level than is actually recorded by a light meter. This yields the "square law" curve, the theoretical relationship between measured light level and perceived brightness:

Perceived Light (%) = 100 x square root (Measured Light (%)/100)

Consider this example (courtesy Lutron): At full brightness, the measured light level is 60fc. At the lowest dimmed level, 10 percent perceived light is desired:

- 1 percent measured light (0.6fcd) is perceived as 10 percent (desired result)
- 5 percent measured light (3fcd) is perceived as 22 percent (2x brighter than desired)
- 10 percent measured light (6fcd) is perceived as 32 percent (3x brighter than desired)

Perception of Light Level Reduction

A dimming issue for some applications is at what point in the change in light level will occupants notice the change.

The Lighting Research Center studied the relative threshold for detection of gradual reduction in light levels. Four sessions were conducted. Sessions A and B were conducted in a room with more than twice the light level of Sessions C and D.

The results are shown in Figure 19-1.

The A, B curve shows:

- More than 90 percent of the population would not notice a 10 percent reduction in lumens

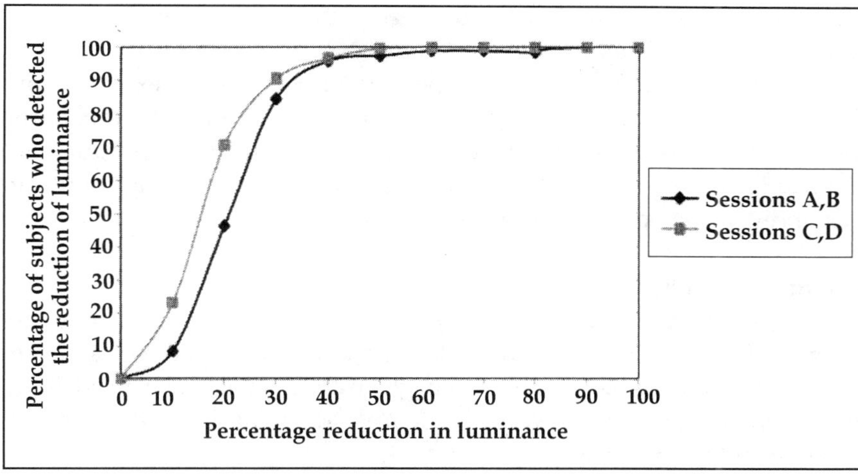

Figure 19-2. Detection of slow light level reduction. *Source: Lighting Research Center*

- About 75 percent would not notice a 15 percent reduction in lumens

- About 55 percent would not notice a 20 percent reduction in lumens

The Lighting Research Center concluded that since the subjects in the experiment were aware that the light level was about to change, which does not match real world conditions, the experiment results can be considered a maximum.

Power Quality

Total harmonic distortion (THD) has been reported to increase on 0-10VDC dimming ballasts as lamp output decreased (*Specifier Reports: Dimming Electronic Ballasts*, Lighting Research Center, October 1999).

Max. THD of less than 3 percent at full light output, for example, increased to a max. THD less than 25 percent at minimum light output. The increase in THD in turn decreased power factor—to a pronounced degree in some ballasts.

The Lighting Research Center concluded that since THD is a percentage of the fundamental current, a high THD at low fundamental current levels associated with low light output levels may not be a concern, as the actual distorted current is small.

Phase-control ballasts also experience THD, but the extent is unknown; in the 1999 *Specifier Reports,* Advance reported that their ballasts experienced less than 10 percent max. THD at full light output, but claimed that current THD and power factor at minimum light output depends on the control device used as well as the ballast.

Energy Consumption

Dimming ballasted lighting system may require higher wattage to operate than fixed light output systems, and do not experience an even lumens-to-wattage reduction. As an illustration, consider a fixed light output ballast powering two F32T8 lamps (see Table 19-2); the lighting system draws 65W of power.

A 0-10VDC ballast requires higher wattage to operate, and at 3 percent lamp output consumes 19 percent of the full input wattage. A phase-control ballast also requires higher wattage to operate, and at 5 percent lamp output consumes 22 percent of the full input wattage.

Note also that shorter lamps are less energy-efficient than longer lamps in dimming applications; each lamp has two electrodes that require the same amount of heating, but represent a larger percentage of the power consumption for the smaller wattage (shorter-length) lamp.

Table 19-2. Comparison of two 120V fixed light output (2) T8 lamp electronic ballasts from Advance Transformer with a 120V (2) T8 lamp 0-10VDC dimming ballast and a 120V (2) T8 lamp phase-control dimming ballast.

					Ballast Factor		*ANSI System Watts*	
Lamps	*Brand/Model*	*Voltage*	*Starting*	*Interface*	*Max.*	*Min.*	*Max.*	*Min.*
(2) F32T8	Centium ICN-2P32-SC	120V	Instant start	Fixed light output	0.88	NA	59	NA
(2) F32T8	Centium ICN-3P32-SC	120V	Instant start	Fixed light output	1.01	NA	65	NA
(2) F32T8	Mark 7 IZT-2S32-SC	120V	Programmed start	0-10VDC	1.00	0.03	68	13
(2) F32T8	Mark X REZ-2S32-SC	120V	Programmed start	Phase-control	1.00	0.05	68	15

Chapter 20

Dimming of High-Intensity Discharge (HID) Lamps

By Craig DiLouie, Lighting Controls Association

High-intensity discharge (HID) lamp dimming has grown in popularity in recent years. Dimming HID lamps can result in energy savings, peak demand reduction and greater flexibility in multi-use spaces.

Dimming reduces energy costs by reducing the input power to the lighting system. It can be used to reduce peak demand and therefore reduce costly utility demand charges that can be a significant component of the total utility cost. And it offers greater flexibility to adapt spaces to different uses.

HID LAMPS

HID light sources, ranging from 20W to 2000W in size, can be found in numerous applications, from retail to industrial to public spaces. It is estimated that there are more than 105 million HID lamps in operation in the United States. HID lighting systems consume 12 percent of all lighting electricity consumed by the commercial sector, 31 percent in the industrial sector, and 87 percent in all outdoor stationary applications—an average of 17 percent of all electricity consumed by all lighting systems in the United States (see Table 20-1).

HID lamps are similarly constructed in that they feature an arc tube of stress- and heat-resistant material that contains gases, metals and the electrodes. They are identified via the predominant distinctive metals contained in the arc tube: high-pressure sodium (sodium), mercury (mercury) and metal halide (metallic halides).

The arc tube is housed in a protective glass envelope. When starting voltage is applied to the electrodes from the ballast or ignitor, an arc

Table 20-1. Facts and estimates concerning HID usage in the U.S. *Source: U.S. Lighting Market Characterization: National Lighting Inventory and Energy Consumption Estimate, Navigant Consulting, Inc./U.S. Department of Energy, September 2002.*

	Commercial	*Industrial Stationary Residential)*	*Outdoor (Including*	*All*
Estimated number of HID lamps/U.S.	30.9 million	15.2 million	54.9 million	105.4 million
Average number of HID lamps/building	7	67	—	—
Operating hours/day	10.1	13.9	11.3	11
Distribution of HID lamps/sector	2 percent	5 percent	75 percent	2 percent
Distribution of installed wattage/sector	11 percent	30 percent	83 percent	7 percent
Distribution of electricity consumed/sector	12 percent	31 percent	87 percent	17 percent
Distribution of lamp output (Terralumens-hour or trillions of lumens/hour)	3,068	2,320	4,677	10,097

is formed between them. Electrons in the arc stream collide with atoms of vaporized metals. The result of this action is the emission of light energy. Due to the high pressures of HID lamp operation, these wavelengths are concentrated in the visible light spectrum and therefore do not require a phosphor coating as a filter.

Of the three types of HID lighting, high-pressure sodium and metal halide are the most efficacious and offer the best color, limiting mercury's use. Metal halide offers superior color quality with a bright white light, while most high-pressure sodium offer the greatest efficiency at the expense of color with an orangish light.

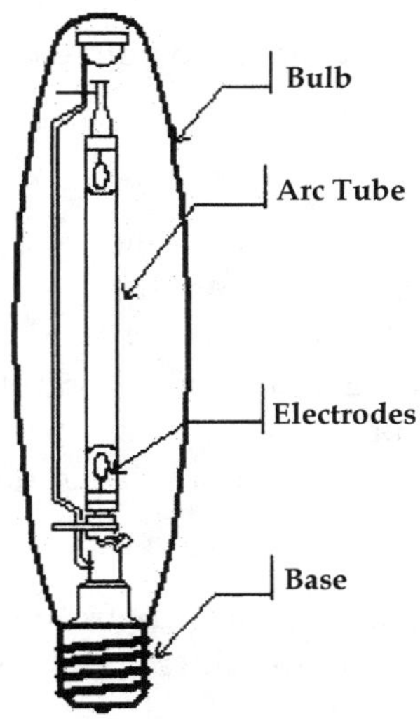

Figure 20-1. High pressure sodium lamp.

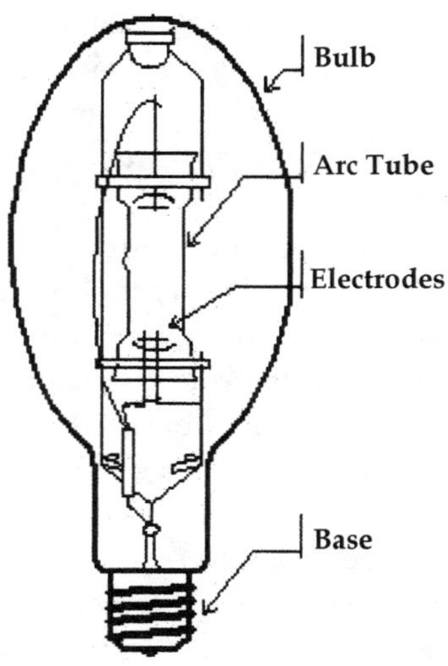

Figure 20-2. Metal halide lamp.

DIMMING STRATEGIES

Dimming can be employed in HID lighting systems to save energy, and enable the space to adapt to different uses, ambient conditions and time of day.

Save Energy

Dimming can be used to save energy during periods when the space is unoccupied but needs to stay lighted for safety and security reasons. Dimming can be achieved either manually via input from a switch or automatically via input from a control device. Automatic dimming can be set to respond to a preset schedule or variable ambient conditions such as occupancy and available daylight.

Occupancy

Dimming is a highly practical control method for saving energy with HID lighting systems to address periods of non-occupancy in spaces that must be constantly lighted.

High pressure sodium lamps can take 3-5 minutes to warm up; they take less than a minute to hot-restrike but don't reach full light for 3-4 minutes. Metal halide lamps take 2-10 minutes to warm up and 12-20 to hot-restrike, while pulse-start metal halide lamps take 1-2 minutes.

Given these characteristics, it is not practical to shut off and restart the lamps based on occupancy if the space must be made usable again quickly. In these situations, the lamps must be operated continuously, resulting in energy waste.

In addition, most lamp manufacturers rate HID lamp life at a minimum of 10 hours per start. Any reduction in burn time per start below this minimum will result in shorter lamp life.

If the lamps are dimmed instead in response to a signal from an occupancy sensor or time-programmable controller indicating the space is unoccupied, significant energy savings can occur during these periods, but the lamps will be able to achieve full light output quickly when the space becomes occupied again.

If occupation of the space is predictable, then timers or other time-programmable controllers may be used to deliver the control signal to dim the lamps. If occupation of the space is not predictable, then occupancy sensors may be used.

Daylight Harvesting

Dimming can be used to adjust light levels based on available daylight via input from a photocell.

Peak Demand Reduction

Dimming can be scheduled using a time-programmable controller during times of peak demand, shaving the facility's peak demand and potentially reducing utility demand charges.

Flexibility

HID lighting systems are fixed output systems, but spaces may require different light levels because they are used for multiple purposes. Dimming makes the lighting system flexible and adaptive to different uses of the space.

A school gym, for example, can be dimmed to provide suitable lighting for sports, social events, maintenance and other uses. A wholesale outlet can be dimmed during maintenance and stocking operations. Spaces can also be dimmed to provide lighting for safety and security.

DIMMING TECHNOLOGIES

HID lamps can be dimmed using step-level or continuous-dimming systems.

Step-level Dimming

Step-level dimming enables wattage reduction, usually at 100 percent and a step between 100 percent and 50 percent of rated power, causing step-level dimming systems to often be called two-level or bi-level dimming systems. However, some systems, often called tri-level dimming systems, can operate at three fixed light levels.

Step-level dimming is ideal for saving energy and providing lighting for safety and security during hours of non-occupancy. Tri-level dimming provides this benefit but offers a greater degree of flexibility to address multiple uses of the space.

This dimming method usually employs a constant-wattage autotransformer (CWA) magnetic ballast with one or two additional capacitors added to the circuit, depending on whether the ballast provides bi- or tri-level dimming. Relay switching of the capacitors results in addi-

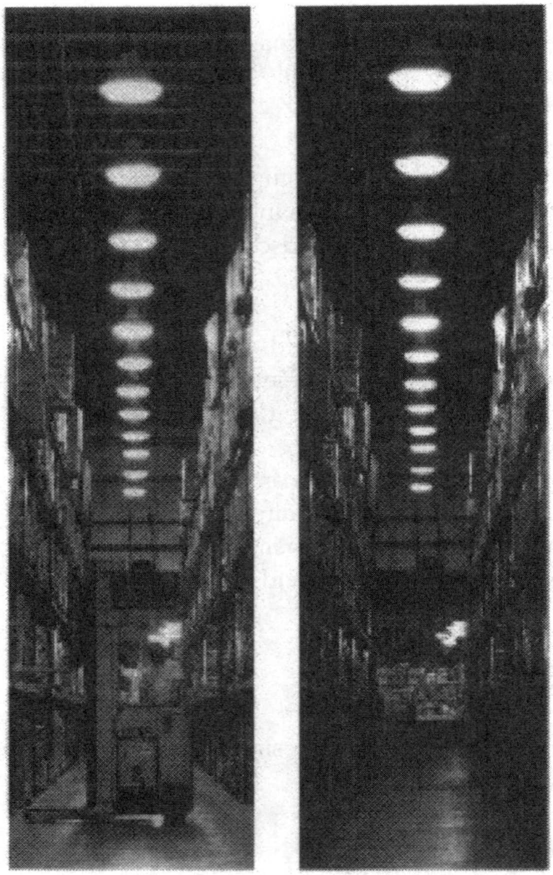

Figure 20-3. Step-dimming energy-saving application in a warehouse. When the space is occupied, the lamps are at full input power and light output (left). When the space is not occupied, an occupancy sensor sends a signal to the dimming system, which dims the lamps while reducing input power (right). *Photo courtesy: Thomas Lighting, Inc.*

tional impedance, which reduces the lamp current and the wattage. The capacitor circuit configuration may be a parallel or series connection.

Step-level dimming is achieved based on input from manual switches, scheduling devices, occupancy sensors and photocells. When the space is occupied, the lamp is brought from its reduced light output to about 80 percent of light output, followed by a brief warm-up time between 80 percent and 100 percent of light output.

Step-level dimming systems using the capacitive-switching method (magnetic dimming ballast) are generally less expensive than continuous dimming systems and are often more cost-effective than HID dimming panels for applications with relatively few fixtures. This type of dimming system also allows individual fixture control. It is suitable for retrofit; in addition, fixtures are available with a dedicated occupancy sensor and dimming ballast, suitable for direct fixture replacement.

Ideal applications for step-dimming include spaces that may be unoccupied for long periods of time but still need to be lighted, such as parking lots, warehouses, supermarkets and malls. High pressure sodium lamps are typically used for parking lots and warehouses, while metal halide lamps are typically used for supermarkets and malls. Step-level dimming systems work with all HID lamp types.

Depending on the lamp type and wattage, in a bi-level dimming system, the Low level may be 15-40 percent of light output and 30-60 percent of wattage. During dimming periods, therefore, energy savings as high as 40-70 percent can result.

A typical application for step-level dimming is a warehouse. When the space is unoccupied—as determined either by an occupancy sensor to detect variable occupancy, an operator with access to a high/low switch, or a timer or other scheduling system—the lamps are dimmed to an energy-saving level. Besides saving energy, the lower light level setting provides minimum lighting for safety and security. During periods of occupancy, the lamps are brought back to full light output.

In outdoor applications such as parking lots, an added bonus of dimming is a reduction in spill light that may impact adjacent properties.

Continuous (Line-voltage) Dimming

A number of technologies are available for smooth, continuous reduction of lamp wattage, including panel-level HID dimming and relatively new electronic HID ballasts. Ideal applications include anywhere it is advantageous to adapt the lighting system to a wide range of light levels to meet various space uses, such as airports, lobbies, classrooms, industrial facilities, sporting arenas, gymnasiums and auditoriums. With the exception of industrial buildings, metal halide lamps are typically used for most of these types of applications. Continuous dimming is also ideal for daylight harvesting by enabling the HID lamp output to be tuned to maintain a constant light level in the space.

Panel-level HID Dimming

This method is used by control systems installed at the electrical panel that reduces the power supplied to the circuit. These control systems accept inputs from occupancy sensors, photocells and time-programmable systems.

The control system may be one of three types:

- Variable-step transformer: Variable-step transformers reduce the voltage supplied to the load, reducing light output and electrical input. They typically operate with existing CWA ballasts. They can reduce rated power down to 50 percent. While they have little impact on power quality, reducing voltage can affect lamp and ballast performance, according to the Lighting Research Center.
- Variable-reactor: This device keeps voltage constant but reduces current, enabling a reduction in rated power down to 30 percent.
- Waveform modification: Also called "wave choppers," these electronic control systems reduce the RMS voltage to the load to reduce rated power down to 50 percent by chopping a part of each voltage cycle. They are used for control of both HID and fluorescent magnetic systems. They are compact and light controls, but can reduce power quality as well as lamp and ballast performance, according to test conducted by the Lighting Research Center. Some devices reduce the light output almost immediately rather than a smooth, gradual reduction, which is perceptible to occupants.

Electronic HID Ballasts

Electronic dimming ballasts for HID lamps are now available in new fixtures and provide continuous dimming, typically from 100-50 percent light output for metal halide and 100-30 percent light output for high pressure sodium lamps so as to preserve lamp life. In addition to dimming, they are designed to operate at a higher efficacy, improved color control, less stroboscopic effect, and harmonic distortion under 20 percent.

Figure 20-4. Electronic HID dimming ballast. *Courtesy: Advance Transformer Co.*

While generally not cost-effective for retrofit, electronic HID ballasts can yield significant energy savings in a new fixture. They are interoperable with occupancy sensors, photocells and time-programmable systems. The signal can be transmitted along the power circuit or low-voltage wires.

Dimming Controls

The dimming signal can be created using one of three types of controls:

- Manual, either local or remote switch
- Automatic, used in conjunction with occupancy sensors or photocells
- Time-programmable, either timers or scheduling systems

Dimming systems can be configured to control a single or multiple zones. The occupancy sensor detects motion and sends a signal to the control system using the power line, low-voltage wire or fiber-optic cable.

RELATED ISSUES

There are a number of technical issues related to dimming HID lamps that lighting professionals should be aware of when specifying an HID dimming system. These issues relate to light output, efficacy, lumen depreciation, service life and color.

Efficacy

The ratio of reduction in wattage to reduction in light output is not proportional with panel-level and step-dimming control systems. Light output will be reduced further than the wattage reduction. In general, light output reductions are about 1.2-1.5 times the power reduction for metal halide lighting systems, and about 1.1-1.4 times the power reduction in high pressure sodium lighting systems. See Table 20-2 for changes in efficacy for a 400W coated metal halide lamp.

Table 20-2. Changes in efficacy for a 400W coated metal halide lamp. Efficacy is defined as the relative light output divided by relative system input power. *Source: Lighting Research Center*

System Input Power (W)	*Relative Efficacy (percent)*
439	100
393	91
354	82
302	79
260	67
247	59

Dimming below 50 Percent

When HID lamps are dimmed below 50 percent of rated power, they may experience degradation in service life, efficacy, color and lumen maintenance, or they may extinguish. Dimming below 50 percent of rated power, in fact, may reduce high pressure sodium and metal halide lamp life by 90 percent. As a result, dimming below 50 percent may void lamp warranties.

NEMA recommends that the maximum recommended dimming level is 50 percent rated lamp wattage for both metal halide and high pressure sodium lamps. NEMA further recommends that the lamps should be operated at full power for at least 15 minutes prior to dimming (unless the lamp is extinguished from a voltage interruption and the input voltage activates the timer, in which case 30 minutes is recommended before dimming.)

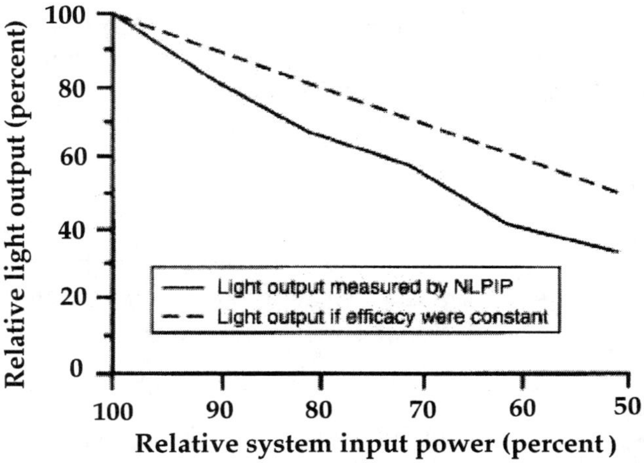

Figure 20-5. Light output versus system input power for a 400W coated metal halide lamp. *Source: Lighting Research Center*

Compatibility

Some panel-level dimming systems are not compatible with electronic ballasts. Self-extinguishing lamps are not recommended for use with dimming systems. Some manufacturers recommend that metal halide lamps be operated base-up to preserve lamp life. Some panel-level dimming systems introduce harmonic currents into the electrical system.

Flicker

Dimming HID lamps, particularly high pressure sodium lamps, can make flicker more visible.

Color

HID lamps can experience a color shift during dimming and also a reduction in color rendering ability. Metal halide lamps are most susceptible to changes in lamp color characteristics.

Clear metal halide lamps, for example, will shift to a higher color temperature or cooler appearance during dimming, from white to blue-green. When a clear metal halide lamp is dimmed to 50 percent of rated power, color temperature can increase 1500K, according to the Lighting Research Center.

Color rendering may also be affected; when a clear metal halide lamp is dimmed to 50 percent of rated power, the Color Rendering Index (CRI) value may decline from 65 to 45.

Coated metal halide lamps experience a much smaller shift and a smaller reduction in CRI than clear lamps.

High pressure sodium lamps can also be affected, typically experiencing a 50-200K reduction in color temperature when they are dimmed, appearing more yellow, while CRI experiences a minimal change.

ALTERNATIVE SOLUTIONS

Facility owners and operators can achieve energy savings with HID lighting without dimming, by considering power reducers, low-wattage HID lamps, and low-bay fluorescent T5 lighting systems.

Power Reducers

Powers reducers, or current limiters, are retrofit devices that can be wired to control an HID ballast or can be installed at the electrical panel to control an entire HID circuit. They are typically designed to work with common CWA ballasts and lamps at least 175W in size. Ideal for overlighted spaces where variable light levels are not needed, they can achieve a preset reduction of 20-25 percent rated power and may extend ballast life by reducing ballast case operating temperature. Reduced-wattage and lower output HID lamps can also be used to retrofit existing fixtures in such applications, as an alternative to power reducers. Although power reducer manufacturers claim that their devices result in little or no reduction in perceived light output, light output will in fact be reduced. It is recommended that lighting professionals conduct a trial installation and measure light levels and wattage before and after installation of the given power reducer.

Fluorescent T5 or T5HO Systems

T5HO lamps have been incorporated into a new type of low-bay (>15 ft.) fixture. This 4- or 6-lamp, instant-on/restrike, high-lumen-maintenance, high-CRI, 20,000- or 28,440-lumen fluorescent fixture has become a popular energy-saving alternative to metal halide in industrial facilities, warehouses, gymnasiums, etc. All things being equal, the T5HO fluorescent is more efficient than metal halide, provides better color rendering and consistency, and has instant-on and instant-restrike, with the trade-off that more lamps and fixtures would be required to light the space, and the fluorescent lamps may not perform as well in cold environments. An interesting side benefit of T5 low-bays is that they can double for emergency lighting.

Chapter 21

Controlling LED Lighting Systems: Introducing the LED Driver*

By Craig DiLouie, Lighting Controls Association

Light-emitting diodes (LEDs) used for illumination are solid-state devices that produce light by passing electric current across layers of semiconductor chips that are housed in a reflector, which in turn is encased in an epoxy lens. The semiconductor material determines the wavelength and subsequent color of the light. The lens converts the LED into a multidirectional or unidirectional light source based on specification.

Colored LEDs currently dominate the exit sign market, with an estimated 85-95 percent of all exit signs sold in the United States using LEDs, and they're making inroads into the traffic signal market, with current penetration estimated at 15-20 percent. They also show significant promise for automobile lighting, and are being sold in a variety of consumer products such as flashlights and light wands. They're also penetrating into mainstream commercial applications such as task lights, accent lights, wall washing, signage, advertising, decorative lighting, display lighting, cove lights and other tight spaces, wall sconces, outdoor/landscape/façade lighting, downlighting and custom lighting.

"Ideal applications today are colored light applications," says Al Marble, Manager—Sales & Market Development for Philips-Advance Transformer. "These are applications where white light sources were previously used and filtered to get the specific color needed. Using color-specific LEDs is cost-effective. The use of LEDs in general lighting applications is still very limited because the quality of white light is still low and also very expensive compared to fluorescent."

The popularity of the light-emitting diode (LED) for a variety of

---

*This chapter originally appeared in *EC&M Magazine*; reprinted here with permission.

lighting fixtures and applications has accelerated in the past year. For example, 44 companies exhibited LED products at Lightfair 2003; this number nearly doubled to 80 companies at this year's Lightfair. LED products won four out of the six top new product awards, including Best of Show.

Why are LEDs becoming so popular? LEDs offer a number of benefits vs. traditional light sources, including:

- Very small size, which increases flexibility to build lumen packages into fixture designs and extends ability to light tight spaces.
- Greater reliability, with no filaments or moving parts; durable and shock-resistant.
- Greater energy efficiency, with 70 percent less energy being consumed.
- Safer and environmentally friendly operation, with less waste and no mercury, and no UV energy and little infrared.
- Color-changing, including the ability to mix colors to generate millions of potential colors.
- Directional light source, which simplifies fixture construction.
- Ability to integrate into architectural materials and to be used to edge-light glass and plastic panels.
- Increased quality, color and strength of light.
- Ability to start at temperatures as low as –40°C.

LEDs are following the major trends in the lighting industry, in which there is strong demand for lighting equipment that is smaller, smarter and more colorful.

"The ideal applications for LEDs are in applications that need colored lighting, compact light sources, and light sources with extremely long life," says Sameer Sodhi, Product Marketing Manager—LED Power Supplies & Controls, OSRAM SYLVANIA, Inc. "LEDs have also reached a point where for long-life applications requiring white light, they are a strong alternative to incandescent lamps."

A new study conducted by the author's firm suggests that engineers consider energy efficiency and long service life to be the most influential attributes in their decision-making to specify LED lighting equipment. According to the U.S. Department of Energy, solid-state lighting has the potential to save enough energy to power the states of Arizona, Colorado and Mississippi and reduce the nation's electric bill by nearly $100 billion over the next 20 years. Architects and lighting designers also consider the small size of the light source and fixtures to be highly influential, and architects further consider rugged operation and ability to change colors to be highly influential.

The study further suggests that engineers are confident about specifying LEDs in the future and see few major barriers to specification, but are more conservative about the use of this technology—less willing to work with new versus traditional suppliers, and most interested in specifying LEDs to replace conventional light sources in traditional fixture types.

"LEDs offer an exciting addition to the world of lighting," says Sodhi. "Not only do they offer a substitute to traditional light sources for certain applications, but also open up a new domain of lighting applications such as decorative architectural lighting."

As engineers become more familiar with LEDs, taking advantage of abundant literature and press coverage, they will need to also familiarize themselves with another component of the LED system that is getting less attention—the LED driver.

LED DRIVER: FUNCTION

LEDs are low-voltage light sources, requiring a constant DC voltage or current to operate optimally. Operating on a low-voltage DC power supply enables LEDs to be easily adaptable to different power supplies, permits longer stand-by power, and increases safety. Individual LEDs used for illumination require 2-4V of direct current (DC) power and several hundred mA of current. As LEDs are connected in series in an array, higher voltage is required.

In addition, during operation, the light source must be protected from line-voltage fluctuations. Changes in voltage can produce a disproportional change in current, which in turn can cause light output to vary, as LED light output is proportional to current and is rated for a current range. If current exceeds the manufacturer recommendations, the LEDs can become brighter, but their light output can degrade at a

Figure 21-1. After an electrical fire destroyed the face of a Carl's Jr. fast-food franchise sign, the neon signage was replaced with new LED signage powered by Advance Transformer's signPRO LED drivers. Input watts dropped from 200W to 38W with the LED system, producing a payback of less than two years. "Our signs are critical to our image and presence. Based on the simplicity of the system, its safety, energy efficiency and ease of installation, LEDs are an optimal solution for our chain," says Jim Sheradin, Manager of Facilities for CKE Restaurants, Carl's Jr.'s parent company.

faster rate due to heat, shortening useful life, which may be defined as the point at which light output declines by 50 percent.

LEDs, therefore, require a device that can convert incoming AC power to the proper DC starting voltage, and regulate the current flowing through the LED during operation. The driver converts 120V (or other voltage) 60Hz AC power to low-voltage DC power required by the LEDs, and protects the LEDs from line-voltage fluctuations.

"An LED driver is the power supply for an LED system, much like a ballast is to a fluorescent or HID lighting system," says Marble.

LED drivers may be constant voltage types (usually 10V, 12V and 24V) or constant current types (350mA, 700mA and 1A). Some drivers are manufactured to operate specific LED devices or arrays, while others can operate most commonly available LEDs. LED drivers are usu-

Figure 21-2. Advance Transformer's signPRO damp location-rated driver for Luxeon LEDs. Luxeon LEDs are offered by LumiLeds Lighting, a joint venture between Philips Lighting Company and Agilent Technologies.

ally compact enough to fit inside a junction box, include isolated Class 2 output for safe handling of the load, operate at high system efficiency, and offer remote operation of the power supply.

DIMMING AND COLOR CHANGING

Drivers can enable dimming and color-changing or sequencing of LEDs. LEDs are easily integrated with circuits to control dimming and color-changing so that these functions can respond to preset commands or occupant presence or commands. Most LED drivers are compatible with commercially available 0-10V control devices and systems such as occupancy sensors, photocells, wallbox dimmers, remote controls, architectural and theatrical controls, and building and lighting automation systems. LEDs can also work with devices governed by the DMX and digital addressable lighting interface (DALI) protocols and, in the future, may include wireless (RF) as a control option.

"With the use of fully electronic drivers, the possibilities are endless," says Marble. "This area is only now being developed, but tighter integration of all electronic components is expected to reduce the use of discrete components in the field and simply application."

Drivers with dimming capability can dim the LED light output the full range from 100 percent to 0 percent. Dimming drivers can dim LEDs by reduction in the forward current, pulse width modulation (PWM) via digital control, or more sophisticated methods. Most dimming drivers operate using the PWM method. With this method, the frequency must be as high as hundreds of thousands of modulations per second so that the LED appears to be continuously lighted without flicker. A benefit of the PWM method is that it enables dimming with minimal color shift in the LED output. According to the Lighting Research Center, dimming causes LEDs to experience a similar shift in spectral power distribution as an incandescent lamp. However, if colored LEDs in an array are used to produce white light, the amount of shift, particularly with red and yellow LEDs, may produce an undesirable effect on the white light that is produced by the system.

Dimming does not result in a loss of efficiency. During dimming, the LEDs are still operated at the same voltage and current as during full light output. In addition, lamp life is not affected by dimming, as is sometimes the case with frequently dimmed fluorescent lighting. Rather, dimming LEDs may lengthen the useful life of LEDs, because dimming can reduce operating temperatures inside the light source.

Drivers can also be used to enable color-changing or sequencing. This can be achieved by dimming a mix of colored LEDs in an array to change colors. Another option is that the driver can work with a color sequencer, which receives the 10V or 24V LED driver output and converts it into three-channel output—usually red, blue and green—that can be mixed to create a wide, dynamic range of colors. When a sequencer is used, it generates a preset sequence, with color changes occurring at a speed determined by the specifier. A third option is for each LED to be individually controlled and programmed by interfacing with DMX digital controller, enabling thousands of LEDs to dynamically dim up or down to create a seemingly infinite spectrum of colors.

SPECIFICATION TIPS

Sodhi points out that a common problem with LED system operation involves overloading the driver. LED drivers are rated for a maximum load that must be paid proper attention.

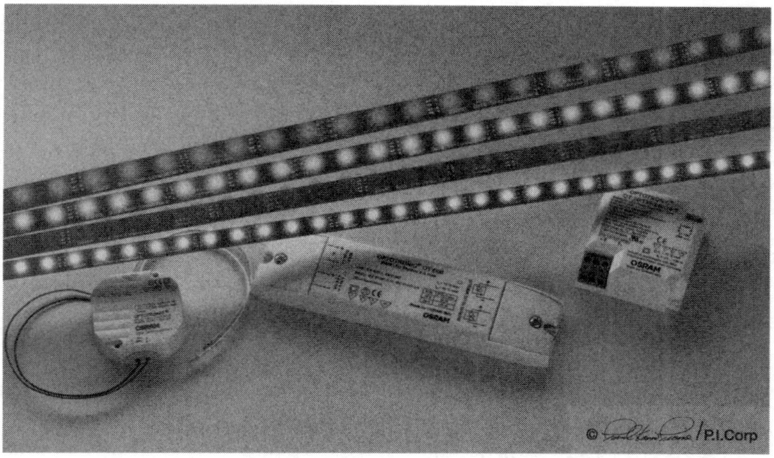

Figure 21-3. LINEARlight Colormix LED Dimmable System from OSRAM SYLVANIA, Inc.

"One of the most common mistakes is to connect too many LED strings in series," he says. "Putting too many strings in series may result in too low a voltage being available to the last string(s) in the chain."

Another common problem, he warns, is using the wrong voltage driver. "When a wrong voltage driver is used, the LEDs will either not light up or may operate at higher currents than intended," he says. "A prudent practice is to check the voltage rating of the LED load being used against the rated output voltage of the driver. For example, using a 12V driver on a 10V LED load could result in significantly shorter life of the module."

Sodhi also believes that one of the most important LED driver features to examine is the quality of the DC output voltage of the driver.

"To maximize the light output from the LEDs without overstressing them requires a constant DC current to be maintained through them," he says.

In addition, he cautions that remote mounting of the driver results in voltage drops and power losses on the DC wiring that must be properly accounted for.

Finally, Sodhi advises specifiers to be aware of ambient temperatures at the application. While LEDs have the ability to start at temperatures as low as –40°C, operating them at cold ambient temperatures can cause operating problems. "LEDs draw higher power at cold ambient temperatures, the opposite of what happens with fluorescent lamps, and this can lead to system malfunction," he warns. "For outdoor applications where the power supply is mounted remotely, the maximum LED load on the driver should be de-rated by 10-20 percent to avoid system conflicts during cold temperatures."

Marble points out that special attention should be paid to the environmental rating of the driver: Most drivers are "dry location only" in type and must be installed in a weatherproof electrical enclosure if used outdoors. Damp location drivers should be used in signs or raceways where some moisture is expected, and wet location drivers are typically supplied in a pre-assembled, sealed enclosure for mounting outdoors.

"Make sure that the driver is rated for use in its environment," he says. "And make sure that the driver has been evaluated and rated for use within the particular LED system."

Marble also believes that UL Class 2 ratings, required for LEDs in sign applications, can benefit general lighting applications.

"UL Class 2 mandates that the driver has voltage, current and power below certain levels on the secondary," he says. UL Class 2 rated LED drivers provide electrical isolation from the AC line voltage, which allows for safe handling of the LEDs being operated at low-level DC voltages.

He also recommends drivers that have short-circuit protection, that are designed specifically for the given application, and that can handle temperature extremes.

"Off-the-shelf DC power supplies are typically designed for room temperature applications such as IT or telecom," he adds. "Such power supplies may operate erratically or not at all under the rigors of a lighting application."

Finally, Marble advises that there are heat issues with LEDs even during normal operation. "LEDs are occasionally and incorrectly believed to generate little or no heat," he says, pointing out that there can be substantial heat generated in higher-wattage LED fixtures. "Hopefully, the integrator/fixture manufacturer designed appropriate heat sinks for the system. Still, allowing ample heat dissipation in the installation is good practice, such as mounting to metal or allowing some ventilation if possible."

Chapter 22

Light Fixtures Get Smart

By Craig DiLouie, Lighting Controls Association

In the 1990s, the energy efficiency trend followed a pattern of integration from components to fixtures. The final vision was to integrate the most efficient lamps, ballasts and lighting control methods into a single fixture. At first, the primary goal was energy efficiency, which later expanded to incorporate facilitywide dimming and occupant dimming, which accelerates savings, provides extraordinary flexibility, and can enhance worker satisfaction and motivation.

This vision has been realized with a generation of "intelligent fixtures" from manufacturers such as Cooper, Lightolier and Ledalite. Each manufacturer chose a distinct product strategy to provide value, resulting in real choices in regards to cost and capabilities for specifiers and owners based on project needs.

All of these products have several common threads. All intelligent fixtures integrate an intelligent dimming ballast that allows programming and control of individual fixtures and connects it to a central or local interface, putting the focus on the fixture instead of the controls system to gain the benefits of intelligent dimming. Depending on the manufacturer, models are available that integrate sensors for occupancy-based dimming or switching as well as daylight and lumen maintenance dimming.

Together, they represent an effective method to achieve flexibility to lower light levels and accommodate changing space needs, increase worker satisfaction by delivering personal control, and reduce energy costs and peak demand charges.

iGEN FROM LIGHTOLIER

In 2001, Lightolier announced its first intelligent fixture, Agili-T, which featured plug-and-play installation, integral sensor technology, multiple optics and a control system that used the existing LAN. At

Lightfair 2003, the company unveiled a line of fixtures that uses the Digital Addressable Lighting Interface (DALI) protocol, called iGEN.

iGEN is available as an option in most Lightolier brands, including Agili-T, Lytespread, Perflyte, Aleron, Spectral Architectural, Eye-Q, Alter Video Teleconference, Paraplus, Vision Smart, Mini-Beam, Coffaire, Wal-Lyter, Walmaster, Lytecel, Calculite CFL Downlights and Pendalyte CFL. iGEN therefore includes linear, recessed, compact and select decorative fixture types and represents more than 700 products able to cover most of the lighting in a typical commercial building. To date, Lightolier has sold more than 10,000 intelligent fixtures.

The iGEN system starts with a digital ballast that is compatible with DALI, an open protocol used to control the operation of ballasts. This enables all of the ballasts in a lighting system to be networked to each other and to control interfaces, such as networked PCs and wallbox controllers. Each ballast to given a unique address in the network so that it can be individually controlled or ganged in groups. For example, various groups of fixtures can be told to dim to different levels according to time of day. Occupants can also control their local lighting at their workstation PCs, nearby wallbox interface, or with hand-held remotes. In addition, iGEN fixtures can talk back, providing energy monitoring capability and maintenance information such as reports of lamp and ballast failures. Other controls can be integrated into the network, as long as they speak DALI. The fixture itself can also be specified with an integrated occupancy sensor to switch or dim based on occupancy, an option that is expanding to more iGEN products.

"Using the standard DALI protocol, Lightolier is not inventing another system for the industry to figure out," said a spokesperson for Lightolier (no longer with the company at the time of writing). "And as DALI installations proliferate, addressable lighting will become the norm rather than the exception. With multiple component manufacturers producing DALI-compliant products, concerns about proprietary solutions vanish. DALI is evolving to the point where virtually any degree of lighting control is possible." He also sees DALI as a step toward integration with other building control systems that use BACnet, LonWorks, EIB and other building control protocols via gateways to make the concept of the intelligent building a viable reality.

While digital lighting networks are often seen as complex,

Lightolier believes iGEN overcomes the complexity barrier by simplifying the system. "By putting the intelligence inside the fixture—using addressable ballasts and integrated sensors—complications regarding component compatibility and complex control wiring have been eliminated," said the spokesperson. "Because the two-wire iGEN communication circuit is located in the same conduit with the line voltage conductors, we can use five-wire modular cables to simultaneously connect the iGEN fixture to both power and digital communication circuits, while assuring solid, error-free connections." Using these cables, iGEN fixtures and controls can be added, removed or relocated without the use of tools.

Lightolier has also assembled a dedicated iGEN project support team that can provide application analysis, project planning, project management, system set-up, user training and technical support.

The primary benefits of iGEN, says Lightolier, are energy savings, personal dimming control, lamp and ballast failure reporting, scalability, and flexibility to accommodate changing space use and lighting needs. "The most important problem solved by iGEN is how we can provide users with the advantages of personal control and the building owner with the resulting energy savings," said the spokesperson. "Our current data shows over 75 percent of commands are to lower light levels, not raise them. In addition, we are providing a platform for the future so that other energy saving strategies can be implemented, such as load shedding or daylight harvesting. We want the owner to know that his investment will continue to grow as the technology grows."

DLS BY COOPER LIGHTING

DLS stands for Digital Lighting System, an option available across seven Cooper brands—Corelite, Fail-Safe, Halo, Metalux, Neo-Ray, Portfolio and Shaper—which enables the fixtures to be tied together in a multi-scene dimming control system via inclusion of an intelligent dimming ballast. This includes linear and compact fluorescent, incandescent and magnetic low-voltage fixtures.

Cooper's strategy was to introduce intelligence into these brands while keeping the specification process simple and focused on the fixture. The specifier selects a fixture with a DLS ballast, then selects

Figure 22-1. Lightolier's Agili-T with integrated occupancy sensor, separate ballasts and optical compartments for uplighting and downlighting, and full DALI addressable control as well as optional wireless I/R dimming control.

control stations and remotes as needed. "With DLS, the control intelligence resides in the ballast so the lighting design process and specification remain with the fixture," says William Johnson, LC, Marketing Manager for Cooper Lighting. "And the system cost with DLS is in the fixture package, so only minimal additional cost is required for the control station(s) and remote(s)."

The control system is comprised of the fixture/ballast, the IR receiver/control station and a hand-held remote, which provide preset and occupant dimming capabilities. It is designed to work out of the box without programming. Each ballast is pre-programmed with five scenes and is set for Zone 1 so that all the fixtures dim as a group, which can be modified through the use of a master hand-held remote called the "Wizard." Occupants can be given the "Sorcerer" or "Ap-

Figure 22-2. Lightolier's Alter IntelligentVision recessed indirect troffer with integrated sensor.

prentice" remotes for personal dimming control.

The fixtures are daisy-chained to the wall- or ceiling-mounted control station(s) using two low-voltage wires. Up to 12 zones and 12 scenes can be programmed, and up to 10 control stations can be used for control of up to 250 ballasts on a single control wire run. Separate zone control (home-run) wiring is not required.

"DLS eliminates complex wiring schemes normally associated with zone wiring," says Johnson. "Scalability comes into play when additional control stations and zone programming are needed. No special control wiring other than the T-tap daisy-chain is needed to add controls or fixtures up to 250 ballasts."

Johnson regards conference rooms, private offices, computer training rooms and other spaces where multi-scene dimming is desirable as ideal applications for DLS. He says that the elimination of separate zone control wiring makes DLS a good value for spaces where multi-scene dimming is usually considered too costly.

"Our DLS brands can all work together with a multi-scene dimming option that's cost-effective, easy to specify, and simple to install," he says.

Cooper has announced two new additions to the DLS option, an occupancy sensor interface and biaxial dimming ballasts. The occupancy sensor interface allows the DLS system to work with any occupancy sensor and switchpack combination. The relay output of the switchpack is wired to the interface. Light levels during occupancy and non-occupancy can be set at the default mode, which is 100 percent occupied, zero percent unoccupied, or re-programmed to meet user's defined light levels.

ERGOLIGHT BY LEDALITE

Ledalite Architectural Products Inc.'s contribution to the intelligent fixtures arena is Ergolight, a direct/indirect optical system designed to provide optimum lighting while saving energy through advanced control.

"Many energy-saving lighting systems result in poor lighting conditions for end-users, negatively impacting their comfort, performance and satisfaction levels," says Mike Wiebe, Marketing Manager for Ledalite. "Ergolight was designed to generate unsurpassed energy and cost savings while maintaining or improving the visual comfort and productivity of end-users."

The Ergolight fixture, incorporating task-oriented (direct) and ambient (indirect) light components, was designed to be able to provide 50 footcandles at the work surface while minimizing glare on computer screens.

"The standard approach to lighting a space is to bathe the entire space with 50 footcandles from wall to wall," says Wiebe. "This can be overkill as most egress areas do not require this level of illumination." Based on this assumption that traditional troffer layouts overlight corridor and egress spaces, Ledalite recommends putting the fixtures over workstations and allowing the indirect component to provide sufficient illumination for egress spaces and corridors.

The result, according to the company, is an up to 50+ percent reduction in number of fixtures required, which can significantly reduce energy costs and overall life-cycle cost—up to 70-80 percent reduction in lighting energy load.

Ergolight is controllable on several levels. The fixture can be centrally controlled using software that also generates real-time en-

Figure 22-3. Ledalite's Ergolight is a direct/indirect fixture that can be centrally controlled by software or locally controlled at the occupant's PC.

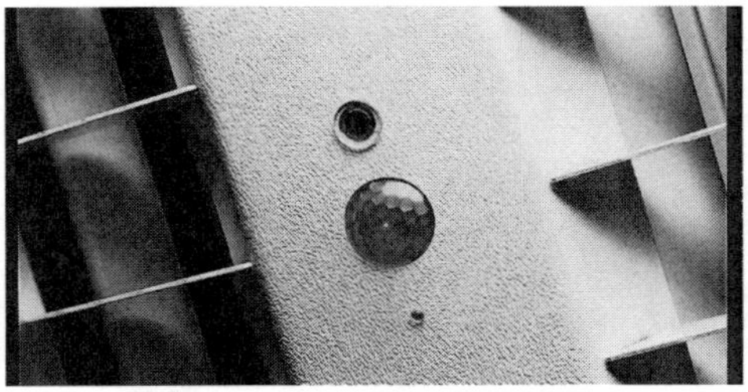

Figure 22-4. Each Ergolight fixture integrates a light sensor for daylight dimming and an occupancy sensor, which gradually dims before turning off for unoccupied spaces.

ergy reports for energy management purposes, and locally controlled at the occupant's PC for personal dimming control. Each fixture integrates a light sensor for daylight dimming and an occupancy sensor, which gradually dims before turning off for unoccupied spaces.

Wiebe says Ergolight uses standard connectors and fits into standard T-bar ceiling grids for simple installation, and that the company designed its software with simple icons and on-screen visual tools.

"Ergolight works well in both retrofit and new construction situations and with a client that is progressive in their thinking," says Wiebe. "It's still not a mainstream product, but it is definitely getting closer to that as time goes by. We believe the demand for quality lighting and reducing energy costs will only continue to grow."

Section V

CASE STUDIES

Chapter 23

Way Station Club House

By the Lighting Controls Association

Location:
Frederick, MD
Mental Health Care Facility Case Study

Architect:
ENSAR Group, Inc. (Gregory Franta, FAIA), Boulder, CO

Lighting Designer:
Clanton & Associates (Nancy Clanton, PE), Boulder, CO

Owners (at time of construction):
Way Station, Inc. (Tena and Grady O'Rear

The architecture of the Way Station Clubhouse directs available daylight to produce an aesthetically pleasing environment that is a critical factor in the healing process. The controlled daylight dimming is an essential part of this integrated lighting system, providing cost-effective and flexible support for the design goals.

Lighting, both daylight and electric, is an integral part of the building's design.

This project set new standards for automated lighting control and facility-wide energy management using daylight with electric light. It provides an outstanding visual environment which supports the healing process. It also demonstrates that energy-efficient buildings that are designed for human comfort are extremely successful. It is quite possible to design both to reduce environmental impact and to construct an affordable commercial building.

"The light really provides a symbol of the kind of openness and positive stance that the organization has taken toward the care of people with serious mental illness," says Tena O'Rear, Owner (at time

Figure 23-1. Way Station Club House.

Figure 23-2. Daylight dimming is an essential part of the integrated lighting system for the Way Station Club House, providing cost-effective and flexible support for the design goals. Shown: Atrium

of construction). "The clients of Way Station love the building. The design is a very open one and from any part of the building its possible to see exactly where you are in relation to the rest of the building. Its a building where people feel a sense of freedom, a sense of lightness... a sense of esteem."

CROSS-SECTION DIAGRAM

Daylight penetration is a vital component of the healing environment in this facility.

Note that almost every interior space has some daylight access, either from the exterior of the building or from the interior courtyard. The roof structures gather the light and direct it to the interior, where light-diffusing banners, light shelves, and reflective surfaces diffuse and moderate any direct glare. This general illumination is supplemented by electric lighting, which only operates when needed.

Figure 23-3. Long view of atrium. Light-diverting cloth panels minimize glare and reflect light into the interior. They also add visual interest to the high space.

Figure 23-4. Office with light shelf. Extensive daylighting minimizes the need for electric lighting. Localized task lighting plus the automatic dimming controls make this a very comfortable work space. The diffuse daylight reflected from the light shelf and the light-colored walls and ceiling also contribute to a feeling of openness and comfort.

DESIGN GOALS

The main goal for this health care facility was to integrate high-quality electric lighting with available daylight, to provide reduction in electric lighting load, a quiet environment, and maximum recuperative benefits from daylight.

Architect

ENSAR Principal, Gregory Franta, brought together all members of the team, including staff, to ensure consensus on design goals, good communication, and that no part of the building was designed in isolation.

Lighting Designer

Primary goals were to balance electric lighting with the daylighting, minimize energy use, especially during peak demand pe-

Figure 23-5. External view of light shelf. The light shelves that shade the south-facing windows reflect light through the window and into the interior. The interior portion of the shelf combines with the exterior to diffuse the light and reflect it deep into the room.

Figure 23-6. Open room with windows. Staff and clients all need daylighting, for health and productivity. Clients with SAD (Seasonal Affective Disorder) especially need lots of daylight. The open design gives a good feeling to visitors, not like the traditional dark and dingy places.

Figure 23-7. Clerestory from the inside. Architecturally integrated clerestory windows bring daylight into the core of the building. Since the clerestory glazing is vertical (like a standard window), they are more weatherproof and easier to keep clean than a typical horizontal skylight. The angle of the entering sunlight is controlled by overhangs and by white cloth banners hanging in the interior. The building is in an historic district, so the roof components are not visible from the street. From the inside, they appear as part of the structural design, and blend in nicely.

riods, and provide a system that worked with the people and their needs. Another unusual goal was to provide a non-institutional feel to the electric lighting system. Buzzing, flickering fluorescents could contribute to negative effects on patients. So one goal was to minimize extraneous noise and light confusion from the electric lighting.

Owners

Way Stations directors, and their staff, wanted a building that was environmentally sound, energy-efficient, and satisfied the needs of their clients. The building had to foster a sense of open communication and well-being, and of harmony with nature.

WHAT WERE THE CONSTRAINTS?

"Money is always a constraint in any project. The owners were very committed to doing the right thing, so keeping controls on the project was simpler. We were looking for good long-term investment in minimizing operating costs."

WHAT WERE THE GREATEST CHALLENGES?

"Since this was one of the first times electric lighting had been dimmed automatically in response to daylight, the commissioning of the system took longer."

WHAT PROMPTED THE DECISION TO USE CONTROLS?

"The building was beautifully daylighted. Each area had daylight coming from multiple directions for balanced light. There was little need for the electric light and it is truly used as an auxiliary system. Therefore, it made sense to dim the electric lighting when not needed to save on energy and to help lessen the mechanical system loading."

SOLUTIONS

Indirect lighting combined with highly reflective surfaces produces a bright interior without compromising visual comfort. The lighting control system must respond to changing daylight levels throughout the day to maintain adequate lighting. Dimming controls provide supplemental electric lighting when daylight levels fall below the preset threshold. As spaces receive more daylight, lights are automatically dimmed. Occupancy sensors provide on/off control for spaces used intermittently. Task lighting provides focused control for small areas.

To achieve the design goals, the Way Station team demonstrated creativity and excellent technical competence.

How did you meet the challenges and constraints?

"Most of the daylight reflects off light shelves and is directed up to the ceiling. The indirect electric lighting, which also lights the ceiling, automatically fills in the light. The luminaires closest to the windows are dimmed depending on the amount of daylight light. As one gets further away from the windows, the electric light gets brighter, filling in for the missing daylight. In smaller areas, occupancy sensors automatically turned the lights on or off. As a result of lighting controls and excellent daylighting design, the mechanical system was downsized from a 100 ton system to a 40 ton system."

What did you learn from doing this project?

"First, it is entirely possible to design affordable commercial buildings which rely on solar energy and energy efficiency to greatly reduce the environmental impact of energy use.

"Second, environmentally benign energy use in buildings is an economic boon. Way Station owners are saving money each year on their building, and they put that savings to work creating jobs in their local economy.

"Third, the designer and owners of the Way Station building have shown that they can create buildings that contribute to environmental well being and personal well-being at the same time.

"The real beauty of Way Station's headquarters is that it is truly a healing place."

What was the worst problem you faced?

"The photosensors for the daylight dimming controls were supposed to be located in the bottom of each indirect luminaire. This did not occur and was missed in the shop drawings. The contractor saw the locations of the light sensors on the plans and assumed it meant ceiling mounted. Therefore, the light sensors ended up directly above the luminaires on the ceiling. When the sensors were operating, the electric lights falsely triggered the sensors so the lights would dim. Then the sensors didn't think there was enough light in the space, so the lights would go up. This 'wave' effect was solved by moving the sensors and recalibrating their sensitivity.

"The glazing on the greenhouse area windows is clear double-pane obscured glass, which allows adequate light without direct solar gain when sun angles are high. The integration of heating and lighting effects in the greenhouse is an excellent example of cooperative design work. Analysis showed that, for example, window performance was more important than wall insulation in saving energy."

BENEFITS

Staff members affirm that the lighting and daylighting systems are dependable, and that the overall feeling of the lighting is natural. The primary goal, to provide a visual environment which enhances healing, has been achieved.

Daylight dimming used in conjunction with indirect lighting resulted in a 41 percent reduction of energy use compared to the same design without daylight dimming control.

Maintenance savings were realized in extended lamp life and reduced maintenance labor needs.

Additional equipment first-costs for this advanced lighting control system were approximately $0.65 per square foot.

There were numerous benefits from the design decisions.

Reduced Energy Use

A reduction in lighting, cooling, and electric by $30,428/yr. or a reduction of 65 percent.

Human Factors

Continuous ventilation system to control air quality and humidity, daylighting for healing mental health patients, extensive plantings in greenhouse and atrium for air quality and food production, and low toxicity materials.

Reduced Construction or Retrofit Costs from Integrated Design

Construction costs were increased by $170,000 for the solar and energy efficient features. Total construction cost was $3,310,000 (or $111/ft.^{2}). This represents a 5 percent increase which provided a 4-year payback.

UNEXPECTED BENEFITS

Environmental and Health Features

Continuous ventilation system to control air quality and humidity

Daylighting

Extensive plantings in greenhouse and atrium

Ceramic tile, low toxicity fabrics and paints

Non- or low-toxic cleaning materials and floor wax used in maintenance

Energy Performance

Percent overall reduction in energy use: 66 percent

Reference Case: 66,100 Btu/ft.^{2}/yr.

Way Station: 22,700 Btu/ft.^{2}/yr.

Auxiliary heating system: Central variable air volume

Solar Features

1028 ft.^{2} greenhouse

2500 ft.^{2} south-facing glass

2-foot (.6 m) exterior and interior light shelves

No west or east glazing

Roof monitors

Skylights with SoLuminaire" daylight trackers

Thermal mass: masonry wall in greenhouse, tile flooring in 80 percent of the building

Energy-efficient Features

R-30 to R-36 ceiling (tapered rigid foam)

R-24 walls: structural block, 2.5 in. foil-faced isocyanurate, exterior brick

Heat Mirror" glazing

High-efficiency lighting equipment and controls

Energy management system

Energy Bills

Energy Bills	*Reference (modeled)*	*Way Station**
Space and water heating	$8,800/yr.	$2,939/yr.
Lighting, cooling, electric	$47,100/yr.	16,672/yr.
Water heating	$2,100/yr.	$734/yr.
Total	$58,000/yr.	$20,345/yr.

*Actual 1992 bills

SPECIFICATIONS & CREDITS

The Way Station project incorporated new design features which required the efforts of many talented people.

Electrical engineer: Engineering Economics, Inc. (John McGovern), Denver, CO.

Controls manufacturer: Lutron Electronics Co., Inc.

Ballast manufacturer: Lutron Electronics Co., Inc.

Luminaire manufacturer: Peerless Lighting Corp.

Photography: Michael Mutmansky

Type of facility: Way Station, Inc. is an organization with a healing mission. The clubhouse is a place where members with long-term mental illness voluntarily come for clinical treatment and rehabilitation. The members also take advantage of socializing with staff and other members.

Size: 2-story building, 30,000 sq. ft.

Completed: February 1991

Chapter 24

Multimedia Classroom, University of Toronto

By the Lighting Controls Association.

The University of Toronto's new "electronic classroom" combines familiar audiovisual equipment, such as slide projectors and VCRs, with such sophisticated equipment as a multi-sync data/video projection system and multi-scene preset dimming controls. Instructors can now electronically enhance their lectures with an integrated user-friendly presentation system.

The Media Center's Electronic Classroom opened in February, 1995. It is located in the Mechanical Engineering Building in a room that has 358 seats.

DESIGN GOALS

The design goal for the electronic classroom was to enhance or improve the learning environment for students and faculty by providing them with an environment equipped with a wide variety of technological options. Classrooms that use a variety of equipment are often not compatible. The room was also designed to be highly intuitive to learn and affordable to purchase and replace components, such as the basic computer that runs it.

SOLUTIONS

To achieve the design goals, the Media Center worked with Adcom Electronics for almost a year to make the Electronic Classroom a reality.

Figure 24-1. Multimedia classroom at the University of Toronto.

They designed a system with preset dimming control in the podium of the Electronic Classroom that let the lecturer select desired combinations of lighting for recall at the touch of a button. Software integrated all devices-VCR, data/video projector, document camera, and lighting. These two systems came together using icons on a touch screen VGA monitor embedded in the console that served as the room's control panel. The instructor presses an on-screen button to activate whatever device is needed; for example a VCR player to introduce a short video clip to supplement, augment, or clarify a topic in a lecture. The lectern design emphasizes practicality and includes multi-task functions that are transparent to the user.

With the system, lighting can be preset to optimum levels; for example, 100 percent for lecture, 50 percent for videoconferencing, 20 percent for data or video viewing, spotlight only for demonstration. Lecturers can concentrate on instruction and, at the touch of a button or screen icon, change lighting levels, audio levels or manipulate other equipment as desired. The system defaults to the original setting when turned off—ready for the next lecturer. Lutron Grafik Eye dimming controls provide four preset room scenes and off for touch button recall.

The first user to test the efficacy of the Electronic Classroom involved a professor in Mechanical Engineering. Positive feedback from both students and the professor led to the decision to conduct may additional sessions from the room. The promising results have led to a host of small activities that have increased the interest on campus. A particularly appealing feature is the ability of instructors, for example, to connect to the network at the Engineering Computing Facility and to bring files from that location to the Electronic Classroom by using an X-terminal. The X-terminal produces workstation-level files and graphics that can be shown by using the high-scan data projector. In other words, professors can extend what they are doing in their labs to the Electronic Classroom. Connection to the Internet is also available from the lectern.

Instructors are able to use either Macintosh or IBM compatible computers in the classroom. The podium has been designed with ports that accommodate both kinds of computers. This means that instructors can use their own notebook computers, for example, to prepare and store their simulations or presentations and then use those same machines during class. All that is required is to plug their computer into the appropriate podium port and then use the ouch screen monitor in the lectern to share their materials with their audience.

The room was equipped with a 486 PC with a touch screen VGA monitor (in 1994). Adcom Electronics' iRoom software managed the room's utilities through the 486 and Microsoft Windows. The iRoom software integrated all the devices (e.g. VCR, high scan data/video projector, document camera, lighting, etc.) using the RS232 connectivity. Using 486 computers and Microsoft software allowed the systems to be replaced inexpensively. The Media Center added 16 more electronic classrooms on campus with laserdisks and networking for videoconferencing capability.

Classes taught in the facility include Mechanical Engineering, Chemical Engineering, Zoology, Chemistry, Mathematics, Business and Management.

BENEFITS

The main benefit of the design decisions was having a classroom that enabled instructors to use sophisticated electronic equipment with a simple interface. The facility was so well received by students and professors that a host of new activities were featured in the electronic classroom. Professors can now extend what they are doing in their labs to the Electronic Classroom.

This educational vision of the future was developed through the efforts of University Information Commons, the lighting controls company, and the technology company personnel. Simplicity in design and usage, budget, and purpose were key elements that were delineated and satisfied. The lighting control system was specified because it is sophisticated, yet simple and easy to use.

The plug-and-play automated characteristics of the room means that the set-up time for instructors to use the technology is quick, and the need to know how to connect different pieces of technology is limited. The projector c an be programmed up to 99 settings which allows for a great deal of flexibility in using it. It means that any type of computer or video source can be connected to it once the settings are programmed. The room is self-sufficient in the sense that the instructor, when trained, operates independently without the need of a technician being present. The instructors are able to operate the room on their own.

The room is easy to use from a technical standpoint and feedback from the students and faculty has been positive, especially in terms of the quality and variety of the audio and visual enhancements that can be inserted into a presentation. However, users have learned that developing new materials such as computer demonstrations, slides, and videos requires a significant amount of time.

SPECIFICATIONS & CREDITS

Owners: University of Toronto, University Information Commons
Technology consultants: Adcom Electronics Ltd. of Toronto R&D Center
Control manufacturer: Lutron Electronics Co., Inc.
Equipment providers: Automated Imaging

Chapter 25

Wal-Mart, City of Industry, CA

By the Lighting Controls Association

Architect:
BSW International (Dru Meadows, RA and Charles Bell), Tulsa, OK

Lighting designer:
Clanton & Associates (Nancy Clanton, P.E.), Boulder, CO

Project management/Efficiency and sustainability studies:
Southern California Edison (Gregg Ander, AIA, and Carlos Haiad), San Dimas, CA

Wal-Mart's new lighting design strategy is intended to increase sales and decrease operating costs, while at the same time reinforcing

Figure 25-1. Front entrance. 288 photovoltaic panels mounted in the front awning provide energy savings and are tied into the grid, eliminating the need for batteries. "From the front entrance, you don't see the extensive array of skylights. It's only when you get inside that you get the full effect of the building's design," says the design team.

Figure 25-2. The Eco-Room. Extensive daylight penetration is incorporated in the design of this retail facility.

the image of the store as a place that cuts overhead in order to offer more competitive pricing.

The City of Industry Wal-Mart is one of their Environmental Demonstration stores. In addition to the usual merchandise, it also has an "Eco-Room," an interactive environmental display area to teach about sustainable design. The building's lighting is an important part of that demonstration.

"[People considering lighting design] definitely need to continue in this particular mode," said Cherie Debrow, Green Coordinator for Wal-Mart. "It can really be a win-win situation, and I think that the more we use these ideas, the more the price comes down. That's been a big concern - that you pay more for recycled or ecology-minded items. We're finding that the tide is shifting, and it's because more designers are taking advantage of it and more building owners are willing to take a chance on it."

DESIGN GOALS

The main goal for this building was to demonstrate an integrated building design that was both environmentally responsible and that exceeded the current building energy standard by at least 25 percent. Wal-Mart wanted this Environmental Demonstration Store to educate the public by example on the benefits of energy efficient technologies and environmental issues. Daylight penetration is a vital component of this facility.

Lighting Designer

"Using the Fresnel lens skylights gave a significant improvement over typical skylights. Using continuous dimming controls was a critical factor in the lighting design. Another goal was to show that lower nighttime light levels are more comfortable than the standard level."

Owners

"Energy savings should exceed current building energy standards, and the building should be an effective educational tool, demonstrating sustainable design."

Project manager

"In addition to all of the other goals mentioned, this is an opportunity to perform long-term monitoring and verification of the system's performance."

Figure 25-3. Store interior. Natural light, supplemented as necessary with electric light, shows off the merchandise to best advantage.

WHAT WERE THE CONSTRAINTS?

"Initial costs were the greatest constraint. Wal-Mart was receptive to the idea of using controlled electric lighting, but we had to prove it's cost-effective and would work for that kind of space."

WHAT WERE THE GREATEST CHALLENGES?

"There were some initial tuning problems with the sensors and controls. They had to be adjusted for unexpected differences in light levels in different areas of the store. Figuring the correct lighting distribution and calculating the cooling load savings from the electric lighting reduction was not simple and required computer simulation."

What prompted the decision to use controls?

"We wanted to minimize operating costs, and at the same time demonstrate that this can be done with no loss of lighting quality."

Was there a "champion" for the use of controls?

Figure 25-4. Skylights over merchandise area. Numerous large skylights enhance the visual environment by providing natural light without glare. The fluorescent lights are automatically dimmed, down to 20 percent, as daylighting increases.

"Southern California Edison was a champion of using controlled lighting from the beginning."

How did you meet the challenges and constraints?

"We had a strong collaborative effort among all the team members. This meant we had an integrated design approach from the start. We used detailed energy simulations and scale modeling during schematic development."

What did you learn from doing this project?

"We did some full-scale testing of night lighting levels, and found that substantially less lighting was needed at night than was normally used."

What were the successful moments or unexpected consequences? What was the worst problem you faced?

"We were surprised at how good it looks. There's no hard data yet, but people seem to stay in the store longer. It feels open and airy, not confined or gloomy. After the initial tuning, the controlled lighting is automatic and virtually maintenance-free."

Figure 25-5. Bank of skylights. The Fresnel lens effect of these skylights gathers and directs lots of daylight into the store, but without allowing direct glare. Thus the interior lighting always appears uniform, without harsh contrast.

Figure 25-6. Flat, rectangular skylights. Three types of skylight are being used, to test their effectiveness: Flat rectangular skylights over general merchandise.

What components did you select, and why?

"The equipment was chosen in response to several factors: the desire to run the luminaires parallel to the front of the store, the requirement that the luminaire provide a good amount of indirect lighting, lamp shielding and cost.

"The lighting installed has a 2-lamp cross section, with 90 percent downlight and 10 percent uplight. In addition to pendant mounted luminaires, luminaires with an asymmetric parabolic reflector on the perimeter walls are used to give visual cues to the boundaries of the space. The dimming system responds automatically to daylight levels."

BENEFITS

These were some of the benefits incurred in this project:

Reduced Energy Use

Lighting energy was reduced 47 percent and total energy reduction was 49 percent compared to California energy standards (Title-24) for 24-hour operation. Payback is estimated at less than three and a half years (excluding the photovoltaic panels).

Reduced Toxic Waste

The low-mercury T8 lamps dramatically reduce pollution, lowering both the total volume and the toxicity of the waste. And, since they pass the EPA test, Wal-Mart estimates the savings in hazardous waste transportation and disposal costs will exceed $5,000 every three to four years, with no reduction in lighting quality. Nearly two million pounds of pollutants are avoided each year as a result of this project.

Reduced Construction or Retrofit Costs from Integrated Design

Removal of a drop ceiling from the design saved about $68,000. Having the entire design team excited and actively involved from the beginning probably minimized some of the usual design problems and delays.

Besides the cost-saving lighting, the store also used non-ozone-depleting refrigerant, and incorporated sustainable and renewable materials wherever possible.

Figure 25-7. Garden area. The lighting quality is excellent throughout the store, at all times of the day or night. And it's safe; even during a daytime power outage, the automatic safety lighting didn't need to come on.

SPECIFICATIONS & CREDITS

Owners: Wal-Mart Stores, Inc.
Daylighting consultant: ENSAR Group, Inc. (Gregory Franta, FAIA), Boulder, CO
Electrical engineer: Consulting Engineers (Jack Vest, III, P.E.), Tulsa, OK
Controls mfg.: Novar Controls Corp.
Ballast Mfg.: Lutron Electronics Co.
Luminaire mfg.: Thomas Industries Inc./Day-Brite
Size: 131,000 sq. ft.
Photography: Michael Mutmansky

Chapter 26

Hyatt Regency, McCormick Place Convention Center, Chicago, IL

Design/Build team:
Mc3D, Inc., Chicago, IL

Architect:
Thompson, Ventulett, Stainback Architects & Associates, Inc., Atlanta, GA

Lighting Designer:
Integrated Lighting Design (Babu Shankar & Christopher Bowsher), Marina Del Rey, CA

The Hyatt Regency McCormick Place is a 32-story, first-class hotel in the heart of Chicago. Architectural dimming controls are used extensively in the lobbies, ballrooms, boardrooms, and restaurant where lighting flexibility is essential to decor and function.
This light shelf provides control of light entering the windows both above the below the shelf.

Decorative lighting was a priority in the design of the public spaces at Hyatt Regency McCormick Place. In public transition and sitting areas such as the reception area and atrium lobby, dimming controls were used to make these spaces appear unique and inviting. In the ballroom and restaurant, dimming is used for mood setting. And, in the boardrooms and conference rooms, dimming provides the flexibility needed to accommodate a variety of presentation media.

Daylight is plentiful in the multiple-story atrium lobby. Light shelves with angled slats are used to control the angle of the light that enters the low glazing and to reflect more daylight up through the higher glazing.

"The most enjoyable part of the building is our restaurant," said Ted Lorenzi, Director of Engineering for the Hyatt. "It's very eclectic,

Figure 26-1. The Hyatt Regency McCormick Place is a 32-story, first-class hotel in the heart of Chicago. Architectural dimming controls are used extensively in the lobbies, ballrooms, boardrooms, and restaurant where lighting flexibility is essential to decor and function.

and it has a lot of bright colors, a lot of different types of lighting, and a lot of different types of light fixtures. It is a kind of multi-purpose room: it's an open area that not only serves as a restaurant, but a bar, and a lounge. Throughout the day the lighting controls really set the whole mood of the area."

DESIGN GOALS

The main design goals for the hotel lighting were to use decorative lighting to enhance the unique appearance of the hotel, to use energy-efficient lighting where appropriate, and to use nighttime fade lighting sparingly. Ease of maintenance was another primary concern. In addition, the lighting designers were constrained by a strict and predetermined budget, and time was limited.

SOLUTIONS

Decorative incandescent lighting was used in public spaces, combined with energy-efficient cove lighting for ambient illumination. Discreet nighttime exterior building lighting was used to call attention to the architectural features instead of floodlighting large areas of the fade. Furthermore, glass and faux alabaster panels were backlighted to customize the reception desk, boardrooms, and corridors.

Lighting controls are standard issue in hotels. They are widely accepted and used for the flexibility they lend to the lighting design. Four-scene dimming control panels were installed in the public areas, such as the reception desk, atrium lobby, ballroom, boardrooms, restaurant and lounge.

Figure 26-2. Close-up of light shelf. This light shelf provides control of light entering the windows both above the below the shelf.

Figure 26-3. Atrium lobby. In the multiple-story atrium lobby, large high windows are shaded by translucent window coverings to control the incoming daylight. Energy-efficient compact fluorescent downlights and fluorescent sconces provide general illumination, while incandescent track lighting provides brightness and sparkle. All of these light sources are controlled by a four-scene preset controller with manual dimming capability.

BENEFITS

Decorative lighting integrated with lighting controls helps set the mood and ambiance of the public spaces. The lighting controls provide the flexibility to adapt the lighting to the time of day. For example, the ambiance of the restaurant is different for breakfast, lunch, dinner, and for the evening. The atrium lobby lighting can be changed in response to the amount of available daylight. The lighting controls lend flexibility to the ballroom and boardrooms, so that they can be adapted to their several potential uses.

Figure 26-4. Registration desk. A variety of lighting techniques was used at the registration desk. First, large faux alabaster panels were backlighted with concealed incandescent and fluorescent sources. Incandescent cove lights are positioned at the top of the wall behind the registration desk to create a decorative play of light on the undulating wall. Recessed incandescent accent lights are positioned over the counter, and task lighting is provided by table lamps on the counter. The lighting in this photograph is controlled by a four-scene preset controller with manual dimming capability.

Figure 26-5. Regency ballroom. In the ballroom, fluorescent cove ambient lighting is accentuated by incandescent downlights and by decorative incandescent pendants. The mood of the ballroom is set by a four-scene preset controller with manual dimming capability.

Figure 26-6. Network restaurants and lounge. Electric lighting contributes to the playful atmosphere of the restaurant and lounge. Dimmable fluorescent cove lighting provides low-level ambient illumination and is integrated into the architecture to enhance the image

of the space. Incandescent downlights and small pendants provide direct light, and decorative sconces behind the bar add to its character. A four-scene preset controller with manual dimming capability provides the flexibility needed to change the ambiance of the restaurant from breakfast through the late evening.

Figure 26-7. Boardroom. The boardroom is lighted using a layered lighting approach. Fluorescent cove lighting brightens the ceiling and provides soft, diffuse ambient illumination. Incandescent accent lighting is provided at the perimeter to highlight the artwork and to create contrast which attracts the eye. Impressive backlighted translucent panels hang over the table and provide direct light. All of these light sources are controlled by a four-scene preset controller with manual dimming capability.

SPECIFICATIONS & CREDITS

Owners: Metropolitan Pier & Exposition Authority, Chicago, Illinois
Controls manufacturer: ALM Systems
Ballast manufacturer: Advance Transformer Co.
Luminaire manufacturer: Lightolier; Winona Lighting
Size: 32-story building
Construction cost: $108 million dollars
Photography: Michael Mutmansky

Chapter 27

New Zoo, Kansas City, MO

Architect:
BNIM Architects (Robert J. Berkebile, FAIA), Kansas City, MO

Lighting Designer:
Clanton & Associates (Nancy Clanton, P.E.), Boulder, CO

Energy Modeling:
ENSAR Group, Inc. (Gregory Franta, FAIA), Boulder, CO

A modern zoo is a far more complex system than the old-fashioned building with stacked cages. Today's zoo visitor expects to learn, not just be entertained. The entry complex to the New Zoo in Kansas City presents an object lesson in sustainable, environmentally aware design. The design team used both daylighting and electric lighting to enhance and support this green agenda.

DESIGN GOALS

The main goal for this building was to produce an environmentally responsible design that would demonstrate that energy efficiency can be elegant and beautiful.

Architect

"The architects aimed for an ecologically balanced environment that would teach visitors about sustainable development."

Lighting Designer

"The goal was to balance daylight and electric light so that the electric light would only supplement the daylight, not duplicate it. Since the daylight was so plentiful, the electric lighting was designed for nighttime social functions, which meant that lower light levels and less uniform light was acceptable."

Figure 27-1. New Zoo in Kansas City, MO.

What were the constraints?

"Money. The construction bid came in much higher than the construction cost estimate, so the project needed to trim 20 percent of the costs. Lighting controls were an easy target, since it didn't cost anything to eliminate them. If controls were more thoroughly integrated with the design from the outset, it would have been more difficult to remove them, and the energy savings would have been substantially greater."

What were the greatest challenges?

"The need to educate the owner, architect, and contractor on the benefits of top-quality lighting equipment and controls. Lighting control technology was relatively unknown at the time of this project (designed 1992-1994, built 1994-1995), so there wasn't a lot of data or experience with controls."

What prompted the decision to use controls?

"Using controls minimizes energy usage, and the philosophy behind the building argued for maximizing daylight use."

Was there a "champion" for the use of controls?

"The architect and lighting designer argued for their use."

Figure 27-2. Lobby. Despite late-stage cost cutting requirements, the integrated design and retention of critical lighting components resulted in a magnificent entry to the zoo.

Figure 27-3. Lobby. The curved gluelam beams and southern pine post-and-beam construction provide a solid structural statement, balanced by the effective combination of daylighting along the whole length and height of the wall, plus post- and ceiling-mounted luminaires for evening and night-time lighting.

Figure 27-4. Lobby. The concrete flooring, timber posts, and even the recycled copper roofing were deliberately chosen for their environmental friendliness, energy efficiency, and subtle hues. The open, high spaces contribute to the perception of the building as connected to the natural environment, not imposed on it.

SOLUTIONS

"Education of the owner and architect was a key issue. Since cost cutting was a major issue, it took a lot of persuasion to keep the good quality lighting equipment on the project. Architectural changes are more costly, so it seemed like an easy fix to eliminate lighting controls to save initial cost. Due to the potential energy savings lost, the architect now regrets the decision to cut the lighting controls."

Figure 27-5. Office area. The work spaces also incorporate natural lighting, which is known to improve productivity and worker satisfaction, and is also healthier. Indirect fluorescent lighting supplements the natural light to provide adequate task lighting. The open design minimizes the need for supplemental lighting.

How did you meet the challenges and constraints?

"We put in a lot of time educating the owner and architect on the value of energy-efficient equipment and controls. We were finally able to keep our top priority of T8 lamps and electronic ballasts."

What did you learn from doing this project?

"Start the education process early!"

What was the worst problem you faced?

"The worst problem was the cost cutting, but the most successful moment was when the owner and architect backed the use of good quality lighting equipment on the project and kept the T8 lamps and electronic ballasts."

BENEFITS

There were numerous benefits of the design decisions.

Reduced Energy Use

The building may not be achieving all of its potential energy savings, but at least the connected load is low compared to other public buildings of its type (0.8 watts/ft.^{2}).

In the design phase, the energy use was predicted to be *78 percent lower than a conventional building* of comparable size and use.

The mechanical (HVAC) system was down-sized because of the excellent glazing specifications and minimal electric lighting loads. This saved money for the total construction.

Human Factors

The building itself is a living example of sustainable design. Visitors notice and enjoy the daylighting, and the educational aspect is very important to the owners.

Reduced Construction or Retrofit Costs from Integrated Design

Minimal electric lighting and excellent glazing resulted in lower HVAC requirements, which saved on initial construction cost.
The best result is that the building is a beautiful example, practicing what it preaches. Visitors can see and feel the effects of designing "green," and the staff have a top-quality work environment.

SPECIFICATIONS & CREDITS

Owners: Friends of the Zoo
Architects: BNIM Architects (Robert J. Berkebile, FAIA; Thompson F. Nelson, FAIA; James C. Tomlinson, AIA; Dale Duncan, RA; David Bell, AIA; Clint Blew; AIA; Dan Maginn, Keith Muller, AIA), Kansas City, MO
Engineers: Structural Engineering Associates, Kansas City, MO; M. E. Group, Kansas City, MO
Controls manufacturer: Sterner Controls
Ballast manufacturer: Advance Transformer Co.
Size: 72,000 sq. ft.

Chapter 28

A Wet Use of Lighting Control

Type of Facility:
Tennis and Volleyball Courts, and Sprinklers

Lighting Representative:
Lurie Systems

Electrical Contractor:
Hatfield Electric

To eliminate time spent by staff checking sprinkler moisture levels at its 10 parks and to cut down water use from leaving sprinklers on too long the Recreation Department of Scottsdale, Arizona, needed a control system.

The department, however, did not turn to traditional plumbing solutions, but instead eyed the control system it used to light its fields, tennis and volleyball courts. Lights illuminating those areas are equipped with button timers that allow light usage for one hour. The lights, are connected to a computer network that automatically disables lights during daylight hours. The system also allows the recreation department to program, operate and monitor the entire parks network from a personal computer in their office.

What made the parks staff turn to the lighting control system for sprinklers was the fact that the system has the capability to control any switch or analog-oriented load. As a result, the department consulted the manufacturer, who studied the unusual request.

DESIGN GOALS

"The Scottsdale Recreation Department wanted one control system to control both the lights and the sprinklers. They also wanted to enable and disable the sprinklers by time-of-day control and switch control. This was made possible by PCI Lighting Control Systems."

WHAT WERE THE CONSTRAINTS?

"To add another system to control the sprinklers would cost the Recreation Department money, not to mention that they would need to learn how the system works. Using one system to control lighting and the sprinklers is simpler and it was already paid for."

WHAT WERE THE GREATEST CHALLENGES?

"Similar applications have been accomplished by PCI Lighting Control Systems, however we have not controlled sprinklers before. We needed to make sure that we had thought of every event possible that would effect the sprinklers and the lighting, and plan accordingly."

What prompted the decision to use controls?

"The staff of the Scottsdale Recreation Department already new and liked the PCI Lighting Control System. They did not need to learn anything new or install any more equipment. Every thing they needed was only keystrokes away."

Was there a champion for the use of controls?

"The staff of Scottsdale Recreation Department realized the potential of the system, like so many of the users of the PCI Lighting Control Systems."

SOLUTIONS

The addition of two relays at each lighting control panel: one for the sprinkler system and one for the switch disable.

The use of one momentary switch input.

Wiring a momentary switch with an "on" wire.

Running that "on" wire through the system's disable relay termination, and then connecting it to the "on" of the sprinkler-switch input channel.

Paralleling the "on" of the sprinkler-switch input channel to the "on" of the adjacent sprinkler disable-switch input channel.

Not connecting the momentary "off" wire.

Programming a 10-minute timer on the sprinkler switch and a second timer to disable the switch after activation.

The revised system also is programmed to run sprinklers on a time-of-day schedule that activates the disable relay when sprinklers need to be turned off. It, in turn, switches the disable relay off when sprinklers need to operate.

The sprinkler schedule can be overridden through the PC, any touch-tone phone (with access code) or by pushing the button timer.

BENEFITS

Staff members affirm that the lighting system is dependable, and that it is easy to control a simple switch, PC command, or touch-tone code can all turn on the lights or sprinklers.

All PCI Lighting Control Systems comes with a computer program that makes its programming even easier. This computer program which we call the Supervisor can issue commands and control your lights or sprinklers from a central computer over an RS-485 network. This Off-Line editor is a big advantage to your system.

An optional Telephone Interface Module allows you to use touch-tone access codes to control your lighting from any touch-tone phone. You can turn the sprinklers on from your own home.

Switches can be assign timers that will automatically turn themselves off after a certain amount of time. Some of our products even allow switch to be prioritized.

SPECIFICATIONS & CREDITS

Controls manufacturer: PCI Lighting Control Systems, Inc.
Size: 10 Parks, containing ball fields, tennis and volleyball courts
Completed: Still Growing

Chapter 29

Other Case Studies

By the National Electrical Manufacturers Association, Lighting Controls Council

SAFEWAY'S CONTROLS PRODUCT FAST PAYBACK

Safeway gained significant benefits when it began to rely on power line-carrier control systems. They turned off rows of fluorescent fixtures in a uniform manner during periods when less light output is sufficient, typically during stocking periods or when they can take advantage of daylight during particularly clear days. Each system consisted of a transmitter mounted on the lighting panelboard and receivers connected to the fixture ballast. The transmitter could be controlled manually, by a time-clock, or by a computer. When less light was acceptable, the transmitter's signals were sent to receivers over existing lighting circuits. No rewiring was needed.

According to the chain's energy management director, the first installation—in a 20,000 square-foot Stockton, California, store—paid for itself within a year.

LIGHTING/PRODUCTIVITY LINK SEEN IN CALIFORNIA

After comprehensive analysis by its own engineers and an energy specialist employed by Pacific Gas & Electric, Control data decided to install all new fluorescent lighting at its Sunnyvale, CA, facility. Particularly affected was the company's 10-member Operations Group, whose work had far-reaching network consequences. Soon after the new lighting was installed, Operations Group personnel reported that the new lighting not only enhanced the appearance of the space, but also improved task visibility. Due particularly to elimination of video display terminal (VDT) screen glare, they were able to boost overall productivity by 6 percent, accomplishing in eight hours what formerly took 8.5.

The value of this benefit was set at $28,000 per year. Even more impressive was the value of downtime avoided each year, estimated to be worth an additional $200,000. Add to the benefits $7,290 worth of operating and maintenance (O&M) costs saved each year (a 60 percent reduction), and the total benefit amounted to $235,290. Given the company's $14,890 investment, simple payback occurred after 23 days, with a simple return on investment (SROI) of 1,580 percent.

NEW CONTROL SYSTEM AT BANK SAVES 25-50 PERCENT IN LIGHTING ENERGY COSTS

A Sacramento, CA bank installed a lighting control system and saved 25-40 percent of lighting energy costs. The system was integrated with a service link that provides on-site intelligence necessary to properly operate the building. The link is custom-programmed for night purge, morning warm-up, ramp and historical optimum start/stop, and remote building monitoring.

Dimming fluorescent lamps slightly during occupancy hours is an important part of the control strategy. If the lights are dimmed 10 percent, electricity is reduced by about 10 percent. Because the lights can be dimmed during periods of high electricity usage, electricity demand charges also are reduced. The service link directs the lights to dim when the building's electricity use is approaching the demand limit.

Further benefits include increased visual comfort because of the ability of the lighting control system to maintain even distribution of light while reducing power to the lamps. Also, the air-conditioning load is lessened, since heat from the fluorescent lamps is reduced.

HOSPITAL TO SAVE $27,000/YEAR WITH NEW FLUORESCENT LIGHTING CONTROL SYSTEM

A Lancaster, PA hospital installed a lighting control system in 40 percent of the patient hallways, the lobby, and the computer information room. A fixed power reduction feature of this system allows personnel to decide how much light is required in different areas of the hospital. A specific level of lighting can be accurately and consistently maintained by setting the fixed lighting levels for the tasks being performed.

The lighting control system contains 10 control modules and 27 output modules. Each control module provides gradual, flicker-free dimming and on/off control for one to six output modules. Each output module provides dimming and on/off control of one 20-ampere branch lighting circuit.

With the installation of the fluorescent lighting control system, an annual savings of $27,000 has been projected.

ULTRASONIC OCCUPANCY SENSORS EXPECTED TO SAVE OVER $400,000 IN OFFICE BUILDINGS

The Koll Company will save more than $400,000 per year by using occupancy sensors to control lighting in its office buildings. The Irvine, CA builder/developer has plans to install the wall- and ceiling-mounted sensors in all of its existing and new high-rise office buildings. Payback is expected in about 18 months.

Occupancy sensors were considered along with other lighting switching systems because energy was being wasted when lighting was on when it should have been off. A typical lighting day for one of Koll's offices is 6 AM to 10 PM. The lights go on to accommodate early arrivals, and stay on until the cleaning crews go through offices late at night. Furthermore, many executives did not spend a lot of time in their offices, making trips for meetings, lunch, coffee, etc.

To reduce the cost of lighting energy, ultrasonic occupancy sensors were used to detect normal minor motions of employees working at their desks and keep the lights on. If the employees leave the room, the lights are automatically switched off after a pre-selected period of time, typically six minutes. Lights can also be turned off manually.

Glossary of Terms

By the National Electrical Manufacturers Association, Lighting Controls Council; and Damon Wood, author of Lighting Upgrades *(The Fairmont Press)*

Ballast: A device that modifies incoming voltage and controls current to provide the electrical conditions necessary to start and operate electric discharge lamps.

Ballast factor: The lumen output of a lamp operated by a commercial ballast divided by the lumen output of the same lamp operated on a reference circuit.

Control group (also control string): A group of lighting fixtures controlled together to provide the basis for comparing the performance of a different group, such as a group with energy-saving lighting controls.

Control zone: All fixtures on one lighting branch circuit.

Daylight: Light from the sky and sun used to provide illumination for the performance of visual tasks.

Daylight (also daylighting) control: An energy-saving lighting control strategy in which a photocell is used with a dimming system to provide a fixed light level at the workplace by increasing the amount of electric light with decreasing daylight levels and decreasing the amount of electric light with increasing daylight.

Dimmer: A control device for varying the light output of lamps.

Direct glare: Glare that is produced by a direct view of light sources. Often the result of insufficiently shielded light sources.

Efficacy: The ratio of light output from a lamp to the electrical input power, expressed in lumens per watt (LPW).

Electronic dimming ballast: A variable output electronic fluorescent ballast.

EMI: Abbreviation for Electromagnetic Interference. High frequency interference (electrical noise) caused by electronic components or fluorescent lamps that interferes with the operation of electrical equipment. EMI is measured in micro-volts, and can be controlled by filters. Because EMI can interfere with communication devices, the Federal Communication Commission (FCC) has established limits for EMI. EMI can also be radiated; see Radio Frequency Interference.

Electronic ballast: A solid-state ballast that uses electronic components to control the lamp at frequencies other than 60 Hz.

Glare: The effect of brightness or differences in brightness within the visual field sufficiently high to cause annoyance, discomfort or loss of visual performance.

Footcandle: The basic measure used to indicate illuminance (level of illumination). One footcandle is equal to one unit of light flux (one lumen) distributed evenly across a one-square-foot surface area.

Illuminance: Lighting level, expressed in footcandles (English unit) or lux (metric unit).

Indirect glare: Glare that is produced from a reflective surface.

Lamp: A light source, commonly called a bulb or tube.

Lighting control: General term referring to electrical devices and techniques necessary to provide the right amount of light where and when needed.

Load shedding: A lighting control strategy for selectively reducing the output of lighting fixtures on a temporary basis as a means to reduce peak demand charges.

Low-voltage switch: A relay (magnetically operated switch) that permits local and remote control of lights, including centralized time clocks or computer control.

Lumen: Basic unit of light flux, or quantity of light.

Lumen maintenance control: an energy-saving lighting control strategy in which a photocell is used with a dimming system to provide a fixed light level over the maintenance cycle.

Luminaire: A complete lighting unit consisting of a lamp (or lamps), together with a housing, the optical components to distribute the light from the lamps, and the electrical components (ballasts, starters, etc.) necessary to operate the lamps. Also called a fixture.

Occupancy sensor: A device that switches lights on and off or dims and brightens them based on the presence or absence of people.

Override: A switch that can be used by occupants to obtain lighting when required outside of normal operating hours. May be activated using a touch-tone telephone.

Photocell: A light-sensitive device for measuring light intensity.

Photometer: An instrument for measuring light intensity and distribution.

Radio frequency interference (RFI): Interference to the radio frequency band caused by other high frequency equipment or devices in the immediate area. Fluorescent lighting systems generate RFI.

Scheduling: An energy-saving lighting control strategy for dimming or otherwise reducing light levels during hours when building space is unoccupied or occupied by individuals with less stringent lighting requirements.

Tuning: An energy-saving lighting control strategy in which the light output of an individual fixture or group of fixtures is adjusted to provide the correct amount of light for a local task.

Index